Components for Evaluation of
Direct-Reading Monitors for Gases and Vapors

DEPARTMENT OF HEALTH AND HUMAN SERVICES
Centers for Disease Control and Prevention
National Institute for Occupational Safety and Health

DISCLAIMER

Mention of any company or product does not constitute endorsement by the National Institute for Occupational Safety and Health (NIOSH). In addition, citations to Web sites external to NIOSH do not constitute NIOSH endorsement of the sponsoring organizations or their programs or products. Furthermore, NIOSH is not responsible for the content of these Web sites. All Web addresses referenced in this document were accessible as of the publication date.

ORDERING INFORMATION

To receive documents or other information about
occupational safety and health topics, contact NIOSH at
Telephone: **1–800–CDC–INFO** (1–800–232–4636)
TTY: 1–888–232–6348
E-mail: cdcinfo@cdc.gov
or visit the NIOSH Web site at **www.cdc.gov/niosh**.
For a monthly update on news at NIOSH, subscribe to
NIOSH eNews by visiting **www.cdc.gov/niosh**/*eNews*.
DHHS (NIOSH) Publication No. 2012–162
July 2012

FOREWORD

The Occupational Safety and Health Act of 1970 (Public Law 91–596) assures, insofar as possible, safe and healthful working conditions for every working man and woman in the Nation. The act charges the National Institute for Occupational Safety and Health (NIOSH) with recommending occupational safety and health standards and describing exposure concentrations that are safe for various periods of employment, including but not limited to the concentrations at which no worker will suffer diminished health, functional capacity, or life expectancy as a result of his or her work experience.

Under that charge and by a 1974 contract, NIOSH and the Occupational Safety and Health Administration jointly undertook the evaluation of sampling and analytical methods for airborne contaminants to determine if current methods met the criterion to produce a result that fell within 25% of the true concentration 95% of the time. In 1995, that protocol was revised.

This document expands the 1995 method development and evaluation experimental testing methods to direct-reading monitors for gases and vapors. It further refines the previous guidelines by applying the most recent research technology and giving additional experimental designs that more fully evaluate monitor performance. These *Components* are provided for laboratory users, consensus standard setting bodies, and manufacturers of direct-reading instrumentation and are compatible with American National Standards Institute/International Society of Automation guidelines. They provide more simplified procedures to estimate the precision, bias, and accuracy of a monitor; to evaluate a monitor relative to the 25% accuracy criterion; and to demonstrate that an atmosphere is relatively *safe*.

John Howard, M.D.
Director
National Institute for Occupational Safety and Health
Centers for Disease Control and Prevention

ABSTRACT

The Occupational Safety and Health Act of 1970 (Public Law 91–596) charged the National Institute for Occupational Safety and Health (NIOSH) with the responsibility for the development and evaluation of sampling and analytical methods for workplace compliance determinations. Under that charge, NIOSH and the Occupational Safety and Health Administration jointly undertook the evaluation of sampling and analytical methods for airborne contaminants by contract in 1974 to determine if methods met the criterion to produce a result falling within 25% of the true concentration 95 times out of 100. The present document further expands the experiments used during the initial methods development and evaluation research to direct-reading monitors for gases and vapors.

This document provides discussion of the physical, operational, and performance characteristics for direct-reading monitors. Guidance is provided for experiments to evaluate response time, calibration, stability, range, limit of measurement, impact of environmental effects, interferences, and reliability of direct-reading monitors. Also included are evaluation criteria for the experiments and details for the calculation of bias, precision and accuracy, and monitor uncertainty.

PREFACE

This *Components* document consists of a main section of three parts, citation of relevant ANSI/ISA standards, and references; eight appendices; and a bibliography. The main section presents background information, reviews various monitor types, and suggests components for monitor evaluation. Appendices A–F assume a prerequisite knowledge of statistics and provide details for the statistical computations, including procedures for calculation of accuracy, bias, precision, alarm set points for alarm system monitors, and explanatory material for relating accuracy to uncertainty. Although all computations can be programmed in spreadsheets, use of a statistical software package is highly recommended. Example computer code is given in Appendices A, B, and G for some of the statistical formulas. For the convenience of interested readers, the more complicated formulas, denoted by bracketed reference numbers, are presented as LaTeX versions in Appendix H.

A companion *Addendum* document, published separately, expands the applicability of the *Components* by presenting methods to be used in evaluating direct-reading monitors for hazard detection in first-responder environments.

Please direct comments, questions, or requests for additional information to the following:
Director, Division of Applied Research and Technology
National Institute for Occupational Safety and Health
4676 Columbia Parkway
Cincinnati, OH 45226-1998
Telephone: 1-513-533-8462 or 1-800-CDC-INFO

CONTENTS

DISCLAIMER ii

ORDERING INFORMATION ii

FOREWORD iii

ABSTRACT iv

PREFACE iv

ABBREVIATIONS ix

DEDICATION xi

ACKNOWLEDGEMENTS xi

Part I. Direct-Reading Monitor Background Information 1

Introduction 1

Definitions 2

    Accuracy 3

    Precision 4

    Bias 4

    Limit of Detection and Limit of Measurement 5

    Measurement Range 5

    Evaluation Range 5

    Interferences 5

    Sampling Rate or Uptake Rate 5

    Exposure Limit 5

    Linearity 6

    Response Time 6

    Sensitivity 6

    Detector 6

    Detector Life 6

    Monitor Uncertainty 7

Part II. Monitor Types 8

Electrochemical Monitors 8

    Conductivity 9

    Potentiometry 10

    Coulometry 10

    Amperometry 11

Ionization 11

    Flame Ionization 11

    Photoionization 13

    Electron Capture 13

Ion Mobility 14

Spectrochemical Monitors 15

    Infrared..15

    Ultraviolet and Visible Light Photometers...16

    Chemiluminescence 17

    Photometric Monitors 17

Thermochemical Monitors 19

    Thermal Conductivity 19

    Heat of Combustion 19

Gas Chromatographs..20

Mass Spectrometers 21

Summary 21

Part III. Suggested Components in Monitor Evaluation **22**

Physical Characteristics—Suggested Documentation 22

    Documentation...22

    Descriptive Information 23

    Physical Information 23

    Portability..23

    Design 23

    Safety 23

Operational Characteristics—Suggested Documentation 24

    Calibration...24

    Ease of Use 24

    Alarm 24

    Power or Battery 24

    Readout 25

    Data Reduction..25

Performance Characteristics 25

    Response Time 25

    Calibration, Linearity, and Drift...26

    Range 27

    Environmental Effects...27

    Precision..27

    Bias 28

    Accuracy 28

    Limit of Measurement...28

    Environmental Interferences...29

    Electromagnetic Interference 29

    Drop and Vibration...29

    Remote Sampling..29

Detector Life 30

Step Change Response and Recovery...30

Supply Voltage Variation...30

Long-Term Stability 30

Monitor Uncertainty..30

Quality System Requirements...30

Reliability...31

Field Evaluation 31

Monitor Results..31

Monitor Evaluation Data Reduction 31

Data Requirements...36

Evaluation and Documentation Reports 36

Relevant Standards..37

References...38

Appendix A. Estimation of Accuracy 41

1. Accuracy 41

2. Estimation 41

3. R Code [R Project 2011] for Equations (A4), (A6), and (A7)...............47

4. References..48

Appendix B. Statistical Evaluation of Bias and Precision for

Individual Monitor Units 49

1. Introduction..49

2. Evaluation of Bias..51

3. Evaluation of Precision 55

4. Evaluation of Precision over Time, Including Day-to-Day Variation...............59

5. Accuracy Calculations 61

6. A Strategy for Obtaining Accuracy Estimates That Take into Account

Varying Bias or S_{rT} by Concentration Level..................................62

7. R Code [R Project 2011] for Section 2.4 63

8. References..64

9. Related Sources...64

Appendix C. Statistical Evaluation of Bias and Precision for a

Population of Monitor Units 65

1. Introduction..65

2. Estimation of Bias..65

3. Estimation of Precision 69

    4. References 71

Appendix D. Measurement Uncertainty **73**
    1. Uncertainty Analysis 73
    2. Comparing Uncertainty and Accuracy 73
    3. References 74

Appendix E. Relationship of the NIOSH Accuracy Criterion to
   Monitor Performance Specifications **75**

Appendix F. Alarm System Evaluation **77**
    1. Introduction 77
    2. Controlling False Negatives (σ and c Known at $C = C_{\mathrm{alarm}}$) 77
    3. σ and c Unknown 77
    4. Controlling False Positives (σ and c Unknown) 79
    5. Verification 80
    6. References 81

Appendix G. R Program for Implementing Appendix F—Lower Confidence
   Limit For Negatives, Controlling False Negatives **83**
    References 83

Appendix H. LaTeX Translations of Selected Mathematical Formulas **85**
    References 85

Bibliography **99**

ABBREVIATIONS

AC	accuracy criterion
ACGIH	American Conference of Governmental Industrial Hygienists
ACS	American Chemical Society
ANOVA	analysis of variance
ANSI	American National Standards Institute
ASTM	ASTM International
B	bias
C	coulomb
°C	degree Celsius
CDC	Centers for Disease Control and Prevention
ceil	ceiling
CEN	European Committee for Standardization
cm	centimeter
DART	Division of Applied Research and Technology
e^-	electron
ECD	electron capture detector
eV	electronvolt
FID	flame ionization detector
FTIR	Fourier transform infrared
g	gram
GC	gas chromatograph
h	hour
hv	energy of a photon (Planck's constant × frequency)
i.d.	internal diameter
IDLH	immediately dangerous to life and health
IEC	International Electrotechnical Commission
IMS	ion mobility spectrometry
IR	infrared
ISA	International Society of Automation (formerly ISA—The Instrumentation, Systems, and Automation Society)
ISO	International Organization for Standardization
J	joule
K	kelvin
kg	kilogram
L	liter
LEL	lower explosive limit
LOD	limit of detection
LOM	limit of measurement
LOQ	limit of quantitation

m	meter
μm	micrometer
min	minute
mL	milliliter
mm	millimeter
mol	mole
MSHA	Mine Safety and Health Administration
ng	nanogram
NIOSH	National Institute for Occupational Safety and Health
nm	nanometer
OSHA	Occupational Safety and Health Administration
PEL	permissible exposure limit
pg	picogram
pH	measure of acidity
PID	photoionization detector
ppb	part per billion
ppm	part per million
S_r	relative standard deviation
S_{rT}	concentration measurement standard deviation relative to true concentration
STEL	short-term exposure limit
TWA	time-weighted average
UV	ultraviolet
VIS	visible

DEDICATION

In memory of Kenneth Busch and Thomas Fischbach.

ACKNOWLEDGEMENTS

These *Components* and the companion *Addendum* were prepared by the staff of the National Institute for Occupational Safety and Health (NIOSH). All contributors are affiliated with NIOSH unless otherwise indicated. We would like to acknowledge the interest and support of the Office of Naval Research in the development of these documents. We extend our thanks to our technical reviewers for their time, constructive comments, and suggestions.

CONTRIBUTORS

Parts I–III
Eugene Kennedy (retired)
Mary Lynn Woebkenberg (retired)
Paul Schlecht (retired)
Eileen Birch
R. Alan Lunsford (retired)
John Snawder
Stanley Shulman

Appendices
Stanley Shulman
Ruiguang Song, CDC
David Bartley (retired)
Amy Feng
Paul Schlecht (retired)
R. Alan Lunsford (retired)

Addendum
Chatten Cowherd, Jr., MRIGlobal
Mary Ann Grelinger, MRIGlobal (retired)
Karin M. Bauer, MRIGlobal
R. Alan Lunsford, (retired)

TECHNICAL REVIEWERS

Frank Dean, Ion Science

Carl Elskamp, OSHA

Martin Harper

Dan Hawley, Brewer Science

Geoff Hewett, American Industrial Hygiene Association, Gas and Vapor Committee

Rodney Hudson, QuickSilver Analytics

Paul Knechtges, U.S. Army

Matthew Magnuson, U.S. Environmental Protection Agency

Susan Moore, OSHA

Martin Petersen (retired)

Robb Pilkington, formerly of University of Missouri, Fire and Rescue Training Institute

Chih-Ming Wang, National Institute of Standards and Technology

EDITORIAL AND PRODUCTION SUPPORT

Document Layout and Format

R. Alan Lunsford (retired)

Brenda Jones

Amy Feng

Yvonne Gagnon

Editorial Review

Anne Votaw (retired)

Dianna Campbell (retired)

Desktop Publishing and Camera Copy Production

Brenda Jones

Web Development

Julie Zimmer

Part I. Direct-Reading Monitor Background Information

Introduction

The Occupational Safety and Health Act of 1970 (Public Law 91–596) charged the National Institute for Occupational Safety and Health (NIOSH) with the responsibility for the development and evaluation of sampling and analytical methods for workplace compliance determinations. Under that charge, NIOSH and the Occupational Safety and Health Administration (OSHA) jointly undertook the evaluation of sampling and analytical methods for airborne contaminants by contract in 1974. During this work, an experimental protocol was developed to define the evaluation criteria for method evaluation [Anderson et al. 1981; Busch and Taylor 1981; NIOSH 1980]. For each method under consideration, the objective of the protocol was to determine if the method would provide results that were within ±25% of the (true) concentration 95% of the time. In 1995, the protocol was revised in *Guidelines for Air Sampling and Analytical Method Development and Evaluation* (*Guidelines*) [NIOSH 1995], based on experience gained in methods development and evaluation research.

The present document, *Components for the Evaluation of Direct-Reading Monitors for Gases and Vapors* (*Components*), further refines the 1995 *Guidelines* and includes an evaluation of direct-reading monitors for gases and vapors. This document does not address passive or diffusive monitors or badges, since there have been many articles published in the technical literature on the use and evaluation of these devices. This new document provides additional experiments and criteria that more

fully and specifically address direct-reading gas and vapor monitor performance and its evaluation.

NIOSH acknowledges that new monitoring technologies other than those described in these *Components* are being developed by the government and commercial sectors. Nevertheless, if new monitoring technology is to be used in occupational settings as described by these *Components*, the principles of evaluation should be very similar. If these experiments are not directly applicable to a monitor under study, then a revised experimental design should be prepared that is appropriate to fully evaluate the monitor. The assistance of a statistician may be required for the preparation of this design. To the maximum extent possible, NIOSH recommends that new monitoring technology be evaluated by using all applicable criteria found in these published *Components*.

These *Components*:

(1) Provide guidance and procedures to estimate the precision, bias, and accuracy of a monitor. As for accuracy, the estimates include the single value that is the best descriptor of the accuracy, and a 90% confidence interval estimate. (Unless explicitly stated otherwise, all confidence interval estimates used in these *Components* are two-sided intervals.)

(2) Provide guidance and procedures to evaluate a monitor relative to the 25% accuracy criterion (or one specified by the user) in terms of one of three mutually exclusive possible conclusions:

- A positive conclusion that there is 95% confidence that the monitor achieves the accuracy criterion. The monitor can be used for both compliance and range finding monitoring.

- A negative conclusion that there is 95% confidence that the monitor fails the accuracy criterion, i.e., that, at best, the method accuracy is worse than 25%. The monitor can only be used for range finding monitoring.

- An inconclusive finding that the monitor does or does not fulfill the accuracy criterion; further research is required to resolve the question. The monitor can, at least, be used for range finding monitoring.

(3) Provide evaluation guidance for direct-reading monitors that need to demonstrate that an atmosphere is relatively *safe*. The most common usage of a *safe* determination involves situations where the monitor shows the concentration to be lower than a recognized occupational exposure limit.

The experiments and definitions used in this document are compatible with those used by the International Society of Automation (ISA, formerly ISA—The Instrumentation, Systems, and Automation Society) to the extent possible. These *Components* are designed for both individual users and the manufacturers of direct-reading monitors. Manufacturers are encouraged to exercise the full range of this document for monitor evaluation if possible, so that consistent evaluation information can be available on the performance of their monitors. End users should be aware of the experiments and criteria provided in this *Components* document so that they can make informed decisions when selecting manufacturer-evaluated monitors for a given monitoring purpose. If some of

the monitors under consideration have not been evaluated, then the *Components* can be used to develop a protocol for this evaluation. The *Components* can be used in part or in whole, depending on the need of the end user and the design of the monitors to be evaluated. For example, when evaluating alarm-only monitors with no readout, experiments dealing with the operating range and limits of measurement of the monitor are not necessarily meaningful. These *Components* can also be used by consensus standard setting bodies for preparation of specific standards for monitor performance.

This document can be used for: (1) selection of the appropriate monitor types for the compounds of interest (Part II—Monitor Types), (2) evaluation of the monitor operation (Part III—Suggested Components in Monitor Evaluation, Physical Characteristics, and Operational Characteristics), (3) evaluation of the monitor performance (Part III—Suggested Components in Monitor Evaluation, Performance Characteristics), and (4) preparation of a technical report on the evaluation (Part III—Suggested Components in Monitor Evaluation, Evaluation and Documentation Reports).

Definitions

This section defines some terms that are used in the rest of this document. Many of these terms are quantities (e.g., **bias**) and the procedures for calculation are listed in Part III—Suggested Components in Monitor Evaluation.

Manufacturers refer to the portable, direct-reading devices that are used to monitor gas and/or vapor levels in the workplace air by a variety of names. The most often used names are **monitor**, **meter**, **detector**, **indicator**, **sen-**

sor, *analyzer*, and *alarm*. In common usage throughout the industry, these names are understood by most users to refer to the whole device. For the sake of uniformity, throughout this document, the term *monitor* will be used to refer to the whole device, while *meter* will refer to only the readout or data display portion of the monitor. *Detector* will refer to the portion of the monitor that actually senses the gas or vapor. *Alarm* will refer only to the audible and/or visual parts of the monitor that are activated whenever a set-point concentration level of gas or vapor is reached or exceeded.

Accuracy

Accuracy is the ability of a monitor to determine the *true* concentration of the environment sampled. Accuracy describes the closeness of a typical measurement to the quantity measured although it is defined and expressed in terms of the relative discrepancy of a typical measurement from the quantity measured. The term *inaccuracy* has also been used interchangeably with the term *accuracy* in the literature [Fowler and Bradley 1990]. In this document, only the term *accuracy* will be used. Accuracy can be a characteristic of a monitor when measurements follow a statistical distribution, such as the normal distribution. Normal distribution is assumed to be useful: it is reasonable as a model for analytical errors—which are measurement errors—even though the distribution of measured environmental concentrations may be log-normal. Unpublished results for the methods studied in Anderson et al. [1981], Busch and Taylor [1981], and NIOSH [1980] indicate that there is little empirical inconsistency with that assumption. Normal theory results are often applicable for other

cases or as good first approximations. Moreover, aside from the relative standard deviation estimates, the analysis is means based. Finally, the authors' unpublished results show that relationships among the method accuracy, precision, and bias that follow from normal theory assumptions hold extremely well for several other distributions, e.g., log-normal, gamma, etc. The special sense of accuracy for a monitor is embodied in the following definition and criterion:

- The accuracy of a monitor is the theoretical maximum error of measurement, expressed as the proportion or percentage of the amount being measured, without regard for the direction of the error, which is achieved with 0.95 probability by the method.

- The accuracy criterion (AC), used in the previous documents [Anderson et al. 1981; Busch and Taylor 1981; NIOSH 1980, 1995] and in this document, requires that a monitor give a result that is within ±25% of the true concentration, having a probability of 95% for an individual observation (i.e., that the accuracy of an acceptable monitor is no greater than 25%).

For a monitor to be accepted as fulfilling the AC, the data from the evaluation study must provide 95% confidence that the accuracy of the monitor is not greater than the AC (25%). To obtain 95% confidence that the accuracy of a monitor satisfies the AC, the 95% confidence limit estimate of the accuracy (see Appendix A) must be less than 25%. For a monitor to be rejected for not meeting the AC, the 5% confidence limit estimate of the accuracy (see Appendix A) must be greater than 25%. If neither of these conditions can be met, the results are inconclusive and more research will be required to reach a definite acceptance or rejection of the monitor.

Precision

Precision is the relative variability of measurements from a homogeneous atmosphere about the mean of the population of measurements. It is calculated by dividing σ, the standard deviation of the measurements, by the measurement mean at a given concentration, designated by μ. The term *imprecision* has also been used interchangeably with the term *precision* in the literature [Fowler and Bradley 1990]. In this document, only the term *precision* will be used. Precision is expressed by the relative standard deviation S_r or the concentration measurement standard deviation relative to the true concentration S_{rT} (see Appendices B and C) of a series of measurements. It reflects the ability of a monitor to replicate measurement results. The statistical definition of the precision is given by

$$S_r = \sigma/\mu, \ S_{rT} = \sigma/C_T.$$

(Note that the 1995 *Guidelines* [NIOSH 1995] defined S_{rT} as a total method precision σ/μ, where σ included contributions from analytical error, sampling error, and pump error. The S_r of the *Guidelines* can denote the relative standard deviation for any of those three components. As will be discussed in Part III of this document, the S_r and S_{rT} used in this document include all known sources of error for the monitor unit.)

If μ represents the true concentration C_T, then S_r is considered the S_{rT}. These *Components* assume that the S_r or S_{rT} of the evaluated monitor is constant or homogeneous over all concentrations tested for the monitor evalua-

tion. This assumption does not imply that the S_r of the monitor is constant over all concentrations, only in those selected for the study. This assumption should be tested using the procedures described in Appendix B, Section 3.1.1.

Bias

Bias B is the relative discrepancy between the mean μ of the distribution of measurements from a monitor and the true concentration being measured C_T, expressed as a fraction. It is given by

$$B = (\mu/C_T) - 1.$$

To meet the AC, acceptable monitors should have a bias magnitude no greater than 10%. Statistical methodology for assessing this is described in Appendices B and C.

Calibrants used also have a small amount of error that is usually defined in the data sheet accompanying the calibrant. It may be expressed as a concentration with a plus/minus value, where the value refers to the uncertainty. The interpretation is that over many calibrant preparations, the bias relative to the designated concentration will be zero, but there can be variability of the actual concentration due to the uncertainty. The uncertainty may be expressed as either a measure of variability or as a multiple of the measure, which should be expressed in the data sheet. In either case, the true concentration is thought to lie between the designated concentration and plus/minus the uncertainty, though the probability with which this statement holds depends on the multiple of the uncertainty that is used.

These *Components* assume that the evaluated monitor bias is constant over all concen-

Components for Evaluation of Direct Reading Monitors

trations tested. The assumption of a constant bias applies only to the range of concentrations tested for the evaluation study and not in general. This assumption should be tested using the procedures described in Appendices B and C for evaluating homogeneity of the bias.

Limit of Detection and Limit of Measurement

Limit of detection (LOD) and limit of measurement (LOM) are defined in *The Automation, Systems, and Instrumentation Dictionary* [ISA 2003]. The ISA LOD is stated to be the smallest value of the measured quantity that produces discernible movement of the monitor indicator. The ISA LOM definition states this value is the smallest value of the measured quantity that can be accurately indicated or recorded.

Measurement Range

Measurement range is the concentration range of the test analyte over which the monitor meets a specified AC.

Evaluation Range

Evaluation range is the range of generated concentrations over which the monitor was evaluated. For most analytes, this range covers concentrations from 0.1 to 2.0 times the exposure limit [Shotwell et al. 1979]. If multiple exposure limits exist, the lowest numerical limit can be used to set the lowest concentration for evaluation. The highest numerical limit can be used to set the upper concentration limit for evaluation. In some cases, this range may be extended to include 10 times the exposure limit [CEN 1994]. In cases where an atmosphere of an analyte was not generated,

the evaluation range can be calculated as the range of concentrations that would be equivalent to the amounts of analyte collected by the monitor for the evaluation experiments, based on typical sampling times and rates.

Interferences

Interferences can be other compounds or conditions that are present with the analyte in the sampled environment that can create difficulties in the determination of the analyte by the monitor. Potential interferences are listed below:

- Compounds that interfere with representative sampling of the analyte.
- Compounds that interfere with accurate analysis of the analyte by the detection system of the monitor.
- Conditions that interfere with representative sampling of the analyte.
- Conditions that interfere with accurate analysis of the analyte by the detection system of the monitor.

Sampling Rate or Uptake Rate

Sampling rate or uptake rate is the volumetric (or equivalent) rate that the air containing the analyte is introduced into the monitor.

Exposure Limit

Exposure limit is the concentration of an analyte above which worker exposure is prohibited or not recommended for a specified period of time during the workday.

For any given analyte, there may be a number of different exposure limits, based on regulations or recommendations from agencies, such as the NIOSH recommended exposure limit (REL) [NIOSH 1992], the OSHA permissible exposure limit (PEL) [29

CFR* 1910.1000], the American Conference of Governmental Industrial Hygienists (ACGIH) Threshold Limit Value (TLV®) [ACGIH 1992], the Mine Safety and Health Administration (MSHA) PEL [30 CFR 56.5001; 57.5001; 71.700 (2003)], etc. These limits also may be international in scope and usually are expressed in one or more of the following terms:

- **Time-weighted average (TWA) concentration:** Concentration measured over a defined time period (e.g., 15 minutes [min], 8 hours [h], 10 h).
- **Short-term exposure limit (STEL):** Time-weighted average concentration measured over a limited sampling period (usually 15 min unless otherwise noted).
- **Ceiling limit (ceil):** Concentration that is not to be exceeded over any time period (e.g., instantaneous to about 5 min).
- **Immediately dangerous to life and health (IDLH):** A situation that poses a threat of exposure to airborne contaminants when that exposure is likely to cause death or immediate or delayed permanent adverse health effects or prevent escape from such an environment [NIOSH 2004]. Failure of an instrument in this situation may contribute to a catastrophic outcome.

Linearity

Linearity is the closeness of a monitor's calibration curve to a mathematically defined straight line.

Response Time

Response time is the time required for a monitor's response to a measurand (e.g., gas, temperature, pressure) to reach a specified fraction

---

*Code of Federal Regulations. See CFR in references.

(some definitions specify 63.4%, but others may be > 90%) of its final response. This lag time depends on the monitor type and measurement conditions. For example, the response time of an amperometric gas sensor for CO depends on both the rate of diffusion of the gas and the electrical time constant of the working electrode.

Sensitivity

Sensitivity is the smallest change in the measured analyte concentration that will produce a reproducible change in a monitor's readout.

Detector

Detector is that part of the direct-reading instrument that *sees* and/or measures and/or quantifies and/or ascertains the dimensions, quantity, or concentration of the gas or vapor of interest. Direct-reading instruments frequently consist of interrelated components that perform a series of functions including sampling, separation, detection, data handling, and readout. Not all direct-reading instruments perform all of these functions, but all direct-reading instruments are capable of detection and have some sort of detector. A detector can be chemical, electrical, mechanical, or physical in nature. A detector may provide a qualitative or quantitative determination of the gas or vapor of interest.

Detector Life

Detector life describes, in general, the time over which a detector can operate within acceptable parameters. As a detector reaches the end of its useful life, its performance degrades beyond acceptable limits. Detector life varies according to the properties of the detector type (e.g., semiconductor, photothermal, optical and fiber optic-based, piezoelectric, pyroelectric, and thermal). The detector manufacturer (or manufacturer whose instrument incorporates a given sensor) should provide an indication of the useful detector life. Whether or not a given detector has passed its useful lifetime can be determined from a quality control graph. When detector response is repeatedly out of control, the user should consider replacing the detector.

Monitor Uncertainty

Monitor uncertainty is the error resulting from the sampling and analysis procedures of a monitor. It can be represented as the mathematical combination of individual errors resulting from the sampling and analysis operations of the monitor [ISO 1993]. (See Appendix D.)

Part II. Monitor Types

Part II presents useful information about commercially available direct-reading monitors for analyzing airborne gases and vapors. The instrumentation discussed provides an on-site indication of the presence of the contaminant(s) of interest and its magnitude in useful units (e.g., part per million, milligram per cubic meter, etc.). Frequently, these monitors use general, nonspecific detectors, but chemispecific detectors are also available.

Direct-reading monitors represent a powerful tool in developing sampling strategies. When correctly used, direct-reading monitors can determine, in real or near-real time, high concentration areas, workers at highest risk, and processes having the highest emissions. Direct-reading instruments, however, are rarely used to monitor compliance with TWA exposure limits. Real time exposure information is useful in solving a variety of gas and vapor exposure problems. This information can guide the hygienist or safety professional in obtaining other more informative and specific samples requiring laboratory analyses.

Direct-reading monitors may be used for area, process, or personal monitoring, and it is convenient to describe three physical classifications for grouping these monitors: (1) *personal monitors* are those monitors small enough to be worn by an individual, (2) *portable monitors* are those easily carried by an individual, and (3) *transportable monitors* are those requiring a cart or other support for movement to or from the monitoring site. Ideally, these monitors operate from self-contained battery power, but many also have or may require, line current.

In this section, the reader will find information on operational, physical, and performance characteristics for each of the monitors described. The monitors are grouped into the following classifications: electrochemical monitors, spectrochemical monitors, thermochemical monitors, gas chromatographic monitors, and mass spectrometers. In each section, the monitor type described is defined in general terms, its principle of detection is explained, and its conditions of application briefly discussed, including capabilities, restrictions, and limitations.

Regardless of the monitor chosen, knowledge of the monitor's capabilities and limitations, as well as the effects of conditions in the proposed monitoring situation, is essential. This knowledge allows for selection of the most appropriate monitor for a given application; it also can translate into more meaningful data results and effective solutions for contaminant control.

It should be recognized that the costs associated with these monitors go beyond the initial purchase price. The costs of supplies and maintenance must also be considered to ensure that the monitor is in optimal working condition. Used properly, direct-reading monitors can conserve resources by allowing more targeted sampling and by reducing the number of laboratory-analyzed samples resulting in "none detected."

Electrochemical Monitors

Electrochemical techniques involve the measurement of electrical signals associated with chemical systems [Strobel and Heineman 1989]. These chemical systems are typically incorporated into electrochemical cells. Electrochemical techniques include monitors that

Components for Evaluation of Direct Reading Monitors

operate on the principles of conductivity, potentiometry, coulometry, and amperometry.

Conductivity

Monitors that measure conductivity rely on the fact that charged species (ions) conduct electricity. Equally significant is the fact that at low concentrations, such as the concentrations typically found in the workplace, conductivity is proportional to concentration. The fundamental equation for conductivity is given by

$$G = \frac{\Lambda C}{1000K},$$

where G is the conductance in siemens (S), Λ is the equivalent conductance in siemens square centimeter per equivalent (S cm^2 equivalent^{-1}), C is the concentration in equivalent per liter (equivalent L^{-1}), and K is the cell constant, a geometric term describing the electrochemical cell in units of cm^{-1}.

As noted in the above equation, conductivity measurement depends on the space between, the area (size) of, and the volume of solution between a pair of electrodes. Conductance is the reciprocal of resistance, that is,

$$G = 1/R,$$

where R is resistance in ohms. The latter is sometimes measured because it is a more fundamental property. Chemicals monitored by conductivity do not need to be in an ionic form in the vapor phase, but may be gases or vapors that form electrolytes by chemical reaction in a liquid, or affect the conducting or semiconducting properties of a solid.

Conductivity measurements are temperature dependent, having a temperature coefficient that can be approximately 2% per degree Celsius (°C) or higher. Monitors that control temperature may use thermostated cabinets; those that compensate for temperature effects do so electronically.

A special case of conductivity instrumentation is one wherein a gold film is used to amalgamate mercury (Hg). In the mercury conductivity detector, the change in resistance of the solid film is measured.

Conductivity is typically a nonspecific technique in that any species ionizable under the given conditions will affect the measurement. The specific conductance λ of each ionizable species is important, because only when the conductivity of interfering electrolytes is either constant and/or negligible can the conductivity of the species of interest be measured.

Several solid-state devices exploit electronic conductivity changes induced in metal oxide semiconductors [Gentry 1993]. Their principle of operation is based on the change in surface conductivity of a semiconductor, such as stannic oxide (SnO_2), as a result of gas adsorption. The adsorbed gas may either directly affect the conductivity or interact with the surface oxygen coverage, which, in turn, affects the conductivity. These monitors are relatively inexpensive, are easy to use, and can be used in oxygen-depleted atmospheres. They are typically used in screening applications and for hazard warning.

Conductivity monitors are primarily used for detection of corrosive gases, e.g., ammonia (NH_3), hydrogen sulfide (H_2S), and sulfur dioxide (SO_2). They are most effectively used in isothermal environments at or near room temperature. Environments having little potential interference are preferred. Chemical prescrubbers can be helpful in removing known interferences.

Potentiometry

Monitors that use a change in electrochemical potential as their principle of detection are most commonly represented by the pH meter. Potentiometry is strictly defined as the measurement of the difference in potential between two electrodes in an electrochemical cell under the condition of zero current. Gases and vapors can react with reagents effecting an oxidation or reduction, the extent of which is proportional to the concentration of the reacting gas. The fundamental equation governing a potentiometric reaction $aA + bB \rightarrow yY + zZ$ is the Nernst equation:

$$E_{cell} = E_{cell}^0 - \frac{RT}{nF} \ln \frac{[Y]^y [Z]^z}{[A]^a [B]^b},$$

where E_{cell} is the cell potential, E_{cell}^0 is the standard cell potential, R is the molar gas constant, T is the temperature, n is the number of electrons involved in the electrode reaction, and F is the Faraday constant.

Although the letters in brackets strictly represent the chemical activities of the reacting species, when a diluted solution is considered, an approximation of the activity using the concentration is reasonable. The equation is simplified at nominal room temperature (25 °C) by converting to the base 10 logarithm and substituting for the constants: $R = 8.314$ J mol^{-1} K^{-1} (joule per mole kelvin), $T = 298$ K, and $F = 96,485$ C mol^{-1} (coulomb per mole). This results in the following equation:

$$E_{cell} = E_{cell}^0 - \frac{0.0591}{n} \log \frac{[Y]^y [Z]^z}{[A]^a [B]^b}.$$

When sampling with a potentiometer, the sampled analyte of interest is most likely represented in the equation by one of the reactants, A or B.

Whereas potentiometry is basically a nonspecific technique, some degree of specificity may be obtained through the selection of the membrane through which the gaseous analyte must diffuse to enter the electrochemical cell, the selection of the reagent, the specific potential range, and the type of electrodes used.

Some potentiometric monitors are diffusion monitors; that is, they do not have a mechanical or electric pump to obtain the sample. They measure a variety of contaminants, including carbon monoxide, chlorine, formaldehyde, hydrogen sulfide, oxides of nitrogen or sulfur, oxygen, and ozone. Preferable application is at, or near, room temperature for area and personal samples (including confined space).

Coulometry

Coulometric monitors use as their principle of detection the determination of the quantity of electricity required to effect the complete electrolysis of the analyte of interest. The amount of electricity required is proportional to the amount of analyte present. This analyte may be the contaminant requiring monitoring, or it may be a chemical with which the contaminant quantitatively reacts. Regardless, the equation governing coulometry is Faraday's:

$$W = \frac{qM}{nF},$$

where W is the mass of substance that is electrolyzed, q is the charge, in coulombs, required to completely electrolyze the substance, M is the formula weight, n is the number of electrons per molecule required for electrolysis, and F is the Faraday constant.

The quantity that a monitor must measure is q. Measurements can be made either directly, by determining the integral (controlled-potential coulometry), or indirectly, by the time required for electrolysis under conditions of constant current (constant-current coulometry). Both approaches work because of the following relationship:

$$q = \int i \, dt,$$

where i is the current in amperes and t is the time in seconds.

Coulometry is free of temperature dependencies. The technique, inherently very accurate, can be nonspecific. Judicious choice of filters, membranes, and electrolytes can be used to improve specificity. The vast majority of these monitors are configured as oxygen or oxygen deficiency monitors although coulometric analyzers are also available for carbon monoxide, chlorine, hydrogen cyanide, hydrogen sulfide, oxides of nitrogen, ozone, and sulfur dioxide. Coulometric detectors can be personal or area monitors and pumped or diffusive samplers.

Amperometry

Amperometric gas sensors are an important class of electrochemical detectors. In these devices, the current generated by reaction of an electroactive species at an electrode is measured at a fixed (or variable) electrode potential applied between the working (sensing) and reference (or counter) electrodes. When operated under diffusion-limited conditions, the sensor current (reaction rate) is proportional to the analyte concentration. The response is typically linear over three orders of magnitude, and measurements with high accuracy and sensitivity [part per million (ppm) and part per billion (ppb)] are possible. Working electrodes are typically constructed of noble metals such as gold and platinum and have high surface

areas. This design provides chemical stability in the sensors' electrolyte solutions and high electrocatalytic activity toward analytes such as CO, H_2S, O_2, Cl_2, and NO [Stetter and Li 2008].

Ionization

There are four types of ionization detectors: flame ionization (FID), photoionization (PID), electron capture (ECD), and ion mobility spectrometry (IMS). Each relies on the ability of its respective energy source (flame, lamp, or radioactivity) to ionize the species of interest.

Flame Ionization

The flame ionization detector (FID) is a robust and easy to operate detector responsive to most organic compounds. It is most responsive to compounds with carbon-hydrogen bonds, especially methane, and is commonly applied to hydrocarbon analysis. In a typical application as a GC detector, effluent from the column flows into a grounded, stainless steel jet at the base of the detector where it is mixed with hydrogen. This fuel-rich mix in the precombustion zone supports a small, hydrogen-air flame at the jet's tip that burns organic compounds, creating positively charged ions and electrons in the flame. To detect the ions formed, a potential difference of a few hundred volts is applied across two electrodes, with the jet also serving as an electrode. A second, cylindrical electrode positioned near the flame serves as the "collector" electrode. Ions attracted to and impacting on the collector induce a current, which is measured by a high impedance circuit (picoammeter and integrator). Data acquisition systems (with analog-to-digital converters) process and display results graphically, with time on the x-axis and total ion on the y-axis.

The FID response has been attributed to a series of reactions that occur from the time analytes enter the precombustion zone of the flame to the point where ions strike the collector electrode. In the fuel-rich, precombustion zone, organic compounds are pyrolyzed to produce single-carbon radical species (CH_3, CH_2, CH, and C). As these species move into the combustion zone of the flame, they react with oxygen to form primary ions and electrons (e^-) through chemical ionization, as exemplified by the following reaction:

$$2\,CH + O_2 \rightarrow 2\,CHO^+ + 2\,e^-.$$

The formed CHO^+ ions are unstable and react rapidly with water in the flame to form hydroxonium ions, H_3O^+, thought to be the primary positive ions in the flame [Hill and Baim 1982]. This reaction sequence is consistent with the detector's response in that single carbon species produced as precursors to ionization are counted, providing a response that is "equal-per-carbon-atom" (e.g., response to one nanomole of hexane is equivalent to two of propane). This reaction sequence, with equilibrium between single-carbon species, also explains the requirement for flame temperature control, which is accomplished through the diluent gas [Hill and Baim 1982]. A chemical ionization reaction producing hydroxonium ions in the flame's combustion zone also is consistent with mass spectrometry findings, where H_3O^+ was detected above the flame during hydrocarbon combustion but CHO^+ was not [Hill and Baim 1982].

The FID has a wide linear range, about 10^6 to 10^7. Though ionization in the detector is not that efficient (about 0.0018% of the molecules produce ions, or about two per 10^5 molecules), it is a very sensitive detector because of the extremely low noise level, about (1 to 2) $\times$ 10^{-12} ampere [Scott 2003]. The minimum detectable mass of n-heptane is about 2×10^{-12} g/sec, which corresponds to a minimum detectable level of about 3×10^{-12} g/mL at a column (GC) flow rate of 20 mL/min. In most applications, the FID is able to detect nanogram (ng) quantities of organic compounds and is excellent in trace analysis. An advantage for GC applications is that the FID is mass sensitive, not concentration sensitive. That is, it responds to mass per unit time rather than mass per unit volume. Thus, response is nearly independent of flow rate and changes in carrier gas flow have little influence on detector response. Further, dilution of the column effluent with hydrogen prior to analyte detection does not reduce the detector response.

The FID detects nearly all carbon-containing compounds, with the exception of a few small molecules such as carbon disulfide (CS_2), carbon tetrachloride (CCl_4), and carbon monoxide (CO). To some extent, it provides a measure of the organic carbon content of the sample, but electronegative atoms such as chlorine and sulfur in organic volatiles depress the response of the carbon atoms to which they are bonded [Holm 1997]. The FID has some selectivity against common air constituents; it does not respond, or responds very little, to water vapor, hydrogen sulfide (H_2S), and ammonia (NH_3). A potential disadvantage of the FID is that analytes are destroyed in the flame and not available for further analysis.

In addition to application as a GC detector, the FID is useful as a portable survey monitor. Small, battery-powered monitors capable of detection at the sub-ppm level are available that allow continuous monitoring up to about 12 h, with an upper range of about 50,000 ppm. Both portable and handheld meters are available. Some models have correction factors for

Components for Evaluation of Direct Reading Monitors

many gases, permitting calibration with methane and sampling for multiple contaminants. Because the detector employs a flame, it may not be safe for use in atmospheres with flammables or combustibles near the lower explosive limit (LEL).

Photoionization

Photoionization is a flameless ionization technique. The contaminant gas or vapor is carried into an ionization chamber where a stable ultraviolet (UV) light source causes the ionization of any species having an ionization potential less than the energy of the incident light [HNU Systems 1986]; that is, photoionization occurs when a molecule absorbs a photon of sufficient energy to cause the molecule to lose an electron and become a positively charged ion:

$$R + h\nu \rightarrow R^+ + e^-,$$

where R is the molecule to be ionized, $h\nu$ is a photon having energy greater than the ionization potential of R, and R^+ is the ionized molecule.

Like the FID, the photoionization detector (PID) responds to ions, but in the PID they are produced by photon absorption rather than chemical ionization. Unlike flame ionization, photoionization is a nondestructive technique. Photoionization detectors have a high-voltage, positive bias electrode to repel the positively charged molecules, accelerating them toward a negatively charged collector electrode. This, in turn, generates a signal at the collector that is proportional to the amount of ionized species. Lamps providing photons of energy up to 10 electronvolt (eV), 10.6 eV, and 11.7 eV are typically offered. PIDs are useful for detection of most volatile organic compounds containing more than two carbon atoms. Higher photon

energy of the UV light source results in more gases detected. Thus, a PID containing an 11.7 eV (argon) lamp will detect formaldehyde and many halogenated compounds not detected by the 10.6 eV (krypton) lamp. While primarily used for detection of organic compounds, the PID has some utility for inorganic compounds, such as hydrogen sulfide, ammonia, and arsine. Constituents of clean air (nitrogen, oxygen, helium), and methane and ethane are not detected. It is necessary to consider if water will interfere, which is dependent on concentration.

Aromatic hydrocarbons exhibit enhanced ionization (and response) in the PID relative to the FID. Photoionization detectors have a linear response with respect to the concentration of a given component. Monitors incorporating PIDs have traditionally been used as area or survey monitors, but personal PIDs are now commercially available. In the past few years, the PID has developed significantly, engaging cell designs that are resistant to contamination effects [Dean and Stockdale 2006]. Sensitivity to benzene at the 50 ppb level is achievable, and the linear dynamic range is considered to be about 10^7.

Electron Capture

The electron capture detector (ECD) is a nondestructive detector that responds strongly to electronegative compounds. It uses a radioactive beta particle (electron) emitter, usually the radionuclide nickel-63 (^{63}Ni), inside a sealed, stainless steel cylinder that is thermostatically controllable, usually from ambient to 375 °C. The original ECD source was tritium absorbed into silver foil, but this source was thermally unstable and soon replaced by the thermally stable ^{63}Ni. A typical source contains a metal foil holding 10 (or 5) millicuries of ^{63}Ni. Because it contains a radioactive source, the ECD is covered under a "General License," requiring a period "wipe test."

In an ECD, electrons from the emitter collide with carrier gas molecules (nitrogen or 5% methane/95% argon), which ionizes the gas molecules to produce a stable cloud of free electrons in the ECD cell. In the absence of organic species, the emitted electrons are attracted to a positively charged anode, generating a constant current. When electronegative compounds (especially chlorinated, fluorinated, or brominated) such as carbon tetrachloride, polychlorinated biphenyls, and halogenated pesticides (e.g., DDT) enter the detector, they immediately combine with (i.e., "capture") some of the free electrons to form negative ions, thereby markedly reducing the current. Electron capture can occur by the following two mechanisms:

$$AB + e^- \rightarrow AB^- + \text{energy}$$

or

$$AB + e^- \rightarrow A^- + B \pm \text{energy},$$

where AB is a reactant. The magnitude of current reduction is a function of both the amount of sample present and its electron affinity. The detector can be operated in two modes, either with a constant potential applied across the cell (DC mode) or in pulsed mode. The ECD response is nonlinear unless the potential across the detector is pulsed, and calibration must be done separately for each sample component that is to be quantified. In pulsed mode, the ECD maintains a constant (standing) current by applying a pulsed potential across the anode and cathode. As the electron population decreases when the electronegative compounds enter the ECD cell from the GC column, the pulse rate increases in proportion to the sample amount. The pulse rate is converted to an analog output.

The ECD is very selective in its response, particularly for halogenated compounds (e.g., pesticides, peroxides, compounds with nitro groups, conjugated carbonyls, and some organometallic compounds); it also is useful for detection of SF_6. It is not sensitive to amines, alcohols, and hydrocarbons. An ECD is highly sensitive, as low as 0.1 picogram (pg), for the compounds it will detect (10 to 1000 times more sensitive than an FID, and a million times more sensitive than a thermal conductivity detector), but it has a limited dynamic range of about 10^3 to 10^4. The typical measurement range is from about 1 ppb to 10 ppm.

Ion Mobility

Of the four ionization techniques, IMS is the newest, dating to the 1970s, and has been somewhat narrowly applied—as, for example, a handheld instrument designed for on-site monitoring of chemical weapons. Ion mobility spectrometry is a method to characterize chemical substances by gas-phase mobilities of ions in a weak electric field. Although many sources can be used, ^{63}Ni foil is the most common ionization source, producing background ions from nitrogen gas:

$$N_2 + \beta^- \rightarrow N_2^+ + \beta'^- + e^-.$$

The N_2^+ begins a series of ion-molecule reactions with the sample in the drift gas (usually N_2 or air), resulting in negative product ions. After the negative product ions are formed, they accelerate in the direction of the weak electric field. The ions also collide with other drift gas molecules. The combination of acceleration and collisions results in a constant average ion velocity $\bar{v}$ that is directly proportional to the electric field E:

$$\bar{v} = KE,$$

where the proportionality constant K is the ion mobility.

The ion mobility is characteristic of the sample. The analytical signal is obtained when the ions strike a conducting flat plate and the resulting current is amplified.

IMS has LODs in the part-per-trillion range. Its selectivity, while a function of ion mobility, allows specificity of detection in the presence of matrix interferences. Since IMS has virtually no time delay between sampling and analysis, real-time monitoring is possible.

All four types of ionization detectors are available in stand-alone monitors, but are also used as detectors in gas chromatographic systems, which will be discussed later.

Spectrochemical Monitors

Monitors having a spectrochemical principle of detection include infrared analyzers, ultraviolet and visible (VIS) light photometers, chemiluminescent detectors, and photometric analyzers [Ingle and Crouch 1988]. In general, spectrochemical analysis uses a spectrum or some portion of a spectrum to determine chemical species. A spectrum is the display of the radiation intensity that is emitted, absorbed, or scattered by a sample. This radiation is related to photon energy via wavelength or frequency.

Infrared

Infrared (IR) spectrometry involves the interaction of the IR portion of the electromagnetic spectrum with matter. Specifically, it is that portion of the spectrum ranging in wavelength from 770 nanometer (nm) to 1000 μm (micrometer), or 12,900 cm^{-1} to 10 cm^{-1} in wavenumber. The IR portion of the spectrum is subdivided into three regions: the near-IR (770 nm to 2.5 μm), the mid-IR (2.5 μm to 50 μm), and the far-IR (50 μm to 1000 μm). The terms near, mid, and far refer to the proximity of the visible portion of the electromagnetic spectrum. IR radiation is not energetic enough to cause electronic transitions in molecules, but it does result in vibrational and rotational transitions. Nearly all molecules absorb IR radiation, making the technique widely applicable. Because the IR spectrum of a given molecular structure is unique to that structure, IR can be fairly specific and useful in compound identification. However, the possibility of overlapping peaks makes any single wavelength of IR measurement in an uncharacterized mixture potentially inaccurate.

IR monitors consist primarily of six major sections: a source of IR radiation, a wavelength selector, a sample cell (closed or open path), appropriate optics, a detector, and a signal processor or readout. Wavelength selection can occur before the sample cell, after the sample cell, or both. In qualitative identification, a complete spectrum may be obtained by IR spectrometry using either a nondispersive or a dispersive technique. A nondispersive IR is a filter photometer, employing interference filters designed to determine a specific pollutant, whereas a dispersive IR uses prisms, gratings, or interferometers to separate radiation into its component wavelengths. Another type of IR instrument relies on Fourier transform (FTIR). Instead of a wavelength selector, the FTIR uses an interferometer to present a phased IR beam to the sample. All wavelength information is collected simultaneously and is averaged over a defined time period to reduce signal noise. The data is then transformed into IR spectral information.

Because it is an absorption technique, infrared spectrometry is governed by Beer's Law:

$$A = \varepsilon bc,$$

where A is the absorbance, ε is the molar absorptivity, b is the path length, and c is the concentration. This equation shows the relationships between the amount of energy absorbed and the length of the path through the sample, and between the absorbed energy and the concentration of the species of interest. The dependency of A on b is significant in discussing parameters of interest because the longer the monitor's path length, the more sensitive the monitor should be. Of significance in Beer's Law, A is $\log(P_o/P)$, where P_o is the original incident radiation, and P is the energy remaining after some is absorbed by the sample. The linear range of IR is limited at any set path length.

An additional monitor parameter of interest is the slit width. The slit width defines the window of energy visible in either the sample or the detector. The width of this slit is proportional to sensitivity and inversely proportional to selectivity and peak resolution.

Monitors, especially area monitors, balance modest precision with selectivity and high throughput. Some monitors are designed to have a fixed wavelength, whereas others are capable of scanning the infrared spectrum. Some of these monitors are designed as general detectors for organics and subgroups, such as hydrocarbons; others are more specific monitors for compounds, such as methane, ethylene, ethane, propane, butane, vehicle emissions, carbon monoxide, carbon dioxide, and several fluorocarbon refrigerants.

A potential disadvantage of infrared monitors is that some ubiquitous compounds, such as water, absorb infrared energy very strongly at certain wavelengths and care must be exercised to avoid making measurements at or near these wavelengths.

Ultraviolet and Visible Light Photometers

Both ultraviolet and visible light photometers operate on the principle of absorption of electromagnetic radiation. The UV is that portion of the electromagnetic spectrum having wavelengths from 10 nm to 350 nm. The actual spectral range for direct-reading UV monitors is closer to 180 nm to 350 nm, which is termed the near-UV, in deference to its proximity to the visible spectrum. The corresponding energy range for the UV is 3.6 eV to 7 eV for the near-UV and 7 eV to 124 eV for the far- or vacuum-UV. The visible spectrum has longer wavelengths than the UV (350 nm to 770 nm) and correspondingly lower energies (1.6 eV to 3.6 eV). As with their IR counterparts, the operational principle (energy absorption) of the UV-VIS monitors is governed by Beer's Law, and the techniques have the same relationships between absorption and concentration, and between absorption and path length. Although the relationship between absorbance and concentration is linear, the value typically measured in spectrophotometry is transmittance T, having a relationship to A given by:

$$A = 2 - \log(T \times 100),$$

where $T \times 100$ is percent transmittance. Transmittance is the ratio of the amount of energy passing through the sample (not absorbed) to the amount of incident energy.

Most UV-VIS monitors are designed to analyze gaseous samples for chemicals, such as ammonia, mercury vapor (which absorbs very strongly at 253.7 nm), oxides of nitrogen, ozone, and sulfur dioxide. A special case of visible spectroscopy is colorimetry, wherein the sample is mixed with a reagent selected to react with the contaminant of interest, forming a colored product. The ability of this colored liquid product to absorb light in the visible region is exploited. This type of monitor, governed by the same chemical principles, can be used as a continuous monitor for a variety of compounds. The UV-VIS monitors are used primarily as area monitors, and are capable of detecting contaminants in the ppm range.

Chemiluminescence

Chemiluminescence is a form of emission spectroscopy wherein spectral information is obtained from nonradiational activation processes [Hodgeson 1974]. A chemical species excited by chemical reactions returns to the lower energy state by emission of a photon. Chemiluminescence is based on the fact that in some chemical reactions, a significant fraction of the intermediates or products are produced in excited electronic states. The emission of photons from these excited electronic states is measured and, if the reaction conditions are arranged appropriately, is proportional to the concentration of the contaminant of interest. Two common chemiluminescence mechanisms are:

$$X + Y \rightarrow I + I^* \rightarrow I + hv \rightarrow P,$$

$$X + Y \rightarrow P + P^* \rightarrow P + hv,$$

where X and Y are reactants, I is an intermediate, and P is the product.

Three conditions must be met to have chemiluminescence take place. First, enough energy must be available to produce the excited state; second, a favorable reaction pathway must be available to produce the excited state;

and third, photon emission must be a favorable deactivation process.

Chemiluminescent monitors analyze gas phase samples, but were developed primarily for oxides of nitrogen and ozone. Because of the chemical reactions involved, the monitors have a high degree of specificity and have a typical LOD on the order of 10 ppb.

Photometric Monitors

This category includes fluorescence analyzers, flame photometric detectors, spectral intensity analyzers, and photometers, primarily reflectance. The first three techniques are all examples of emission spectroscopy, in which the excitation process is radiative; the last category (reflectance) includes automated media advance samplers, branched sequential samplers, and paper tape stain development, all of which use photometric analysis.

Fluorescence

Fluorescence is the emission of photons from molecules in excited states when the excited states are the result of the absorption of energy from some source of radiation. For most molecules, electrons are paired in the lowest energy or ground state. If a molecule absorbs energy from a sufficiently powerful radiation source, such as a mercury or xenon arc lamp, the molecule will become excited, moving an electron to a higher energy state. When the electron returns to the lower, more stable energy condition, it releases the absorbed energy in photons. A significant characteristic of fluorescence is that the emitted radiation is of a longer wavelength (lower energy) than the excitation radiation. An excitation wavelength selector is used to limit the energy to that which will cause fluorescence of the sample while excluding energy wavelengths that may interfere with the detection. The emission wavelength selector isolates the fluorescence peak. Detection is at right angles to allow measurement of the longer wavelength light emitted from the sample while avoiding detection of light from the source, which could cause large errors in measurement. A narrow band of excitation and emission wavelengths can make the monitor very selective and often specific. Typical limits of detection are in the 5 ppb to 10 ppb range.

Flame photometric

Flame photometric detectors can be adjusted to obtain selectivity for ng quantities of sulfur or phosphorus compounds or other compounds. The detector works by measuring the emission of light from a hydrogen flame. Light from the flame impinges upon a mirror and is reflected to an optical filter that allows only light of either 526 μm (for phosphorus) or 394 μm (for sulfur) to pass through to the photomultiplier tube. Other elements can be detected with tuned wavelength filters. Calibration with flame photometric detectors is critical because they exhibit little or no linearity. Typical LODs are in the low ppb range.

Spectral intensity

Spectral intensity monitors measure the radiant power of emission from an analyte because of its nonradiational excitation. Spectral intensity has been used for halide detection by measuring the increased spectral intensity of an alternating current arc (or spark) in the presence of halogenated hydrocarbons. The increased intensity is converted to the concentration of the halogenated compound by using a calibration curve based on the specific compound of interest, as each response curve for each halogenated compound will be different. These monitors have LODs in the tens of ppm range and have limited selectivity, which means that they can differentiate halogenated compounds from nonhalogenated compounds, but cannot differentiate between halogenated compounds.

Photometers (other)

The remaining types of monitors in this category are simply referred to as photometers. The monitors have unique sampling characteristics and detection principles relative to the other spectrochemical monitors, but the principle of operation is based on a spectrum. The majority of these monitors allow for either unattended sampling through automated sampling media advance (i.e., tape samplers, rotating drum samplers, rotating disc samplers, and turntable samplers measuring reflectance) or branched sequential sampling trains. These samplers typically involve a color change of the sampling medium, and the analytic finish is the measurement of the light reflected from the sampling medium. The reflectance monitors can be quite specific through judicious selection of the chemistry for the sampler, and the ability to change the chemistry makes these monitors potentially useful for a wide variety of compounds. These monitors are useful for such toxic species as toluene diisocyanate, ammonia, phosgene, arsine, and hydrogen cyanide.

Another photometer is designed to determine carbon monoxide (CO). This monitor actually measures mercury that is generated via reduction of solid-state mercury oxide. The amount of mercury generated is equal to the quantity of carbon monoxide oxidized in the sample. The mercury is measured with a UV filter photometer.

Other photometer monitors rely on the development of a color stain, wherein the intensity change or the development of the intensity change is measured via a photoelectric cell. This type of monitor is useful primarily for hydrogen sulfide, although one will determine other analytes as a function of the chemically impregnated paper used for color development. All the photometers have LODs in the low ppm range and are very specific for the contaminant(s) of interest.

Thermochemical Monitors

Gases and vapors have certain thermal properties that can be exploited in their analysis [Skoog et al. 1988]. Of the monitors available for industrial hygiene applications, one of two thermal properties—conductivity or heat of combustion—is measured.

Thermal Conductivity

Thermal conductivity detectors are relatively simple devices that operate on the principle that a hot body will lose heat at a rate that depends on the composition of the surrounding gas. That is, the ability of the surrounding gas to conduct heat away from the hot body can be used as a measure of the composition of the gas. In actual practice, a thermal conductivity detector consists of an electrically heated element, or sensing device, whose temperature at constant electrical power depends on the thermal conductivity of the surrounding gas. The resistance of the sensing device is used as a measure of its temperature. Thermal conductivity detectors are universal detectors, responding to all compounds. They have large linear dynamic ranges, on the order of 10^5, and LODs on the order of 10^{-8} gram (g) of solute per milliliter (mL) of carrier gas (10 ppm to 100 ppm for most analytes). Thermal conductivity detectors require consistent temperature and flow control.

Heat of Combustion

Heat of combustion detectors comprise the largest single class of direct-reading monitors for analyzing airborne gases and vapors. They measure the heat released during combustion or reaction of the contaminant gas of interest. The amount of heat released is characteristic of each combustible gas and may be used for quantitative detection.

There are two main mechanisms for operating heat of combustion detectors. The first relies on heated filaments. Upon introduction of the contaminated air into the sample cell, the contaminant comes into contact with a heated source that ignites the contaminant. The resulting heat changes the resistance of the filament. Calibration standards relate the measured change in resistance to the gas concentration. The second mechanism employs catalysts via catalytically heated filaments or oxidation catalysts, and uses one of two methods of detection: a measured resistance change or temperature changes measured via thermocouples or thermistors.

As in thermal conductivity detectors, heat of combustion detectors are nonspecific, universal detectors. Some specificity can be introduced by manipulation of the temperature. That is, the combustion temperature may be controlled so that it is insufficient to combust interfering gases. For monitors using catalysts, some specificity may be introduced by careful selection of the oxidation catalyst.

As the category name implies, heat of combustion detectors are available as generic detectors for combustible gases. Some heat of combustion detectors are more specific for carbon monoxide, ethylene oxide, hydrogen sulfide, methane, and oxygen deficiency. Most of these monitors read out in terms of percent of the lower explosive limit or hundreds of ppm, and the LOD is a function of the analyte of interest.

Gas Chromatographs

In terms of detection of airborne gases and vapors, the detectors used in a gas chromatograph (GC) have, for the most part, been discussed earlier in this section [David 1974; HNU Systems 1986; McNair and Miller 1997]. The most frequently used detectors in GCs designed for industrial hygiene applications are the FID and the PID. The reason gas chromatographs are being discussed separately is threefold:

- There are several direct-reading gas chromatographs commercially available. These field portable gas chromatographs most closely approximate the transfer of laboratory analytical techniques into the field.
- They represent a distinct family of monitors in that they address separation (specificity) and detection in industrial hygiene monitoring.
- They represent one area where a great deal of research and development is ongoing.

In GC analysis, either the sample is injected into the GC by using a gas-tight syringe or the monitor may be capable of obtaining its own sample via a built-in sampling pump. If the sample is a liquid, the monitor must be capable of vaporizing the sample (e.g., using a heated injection port).

The actual separation of the sample into its component parts takes place on the GC column [McNair and Miller 1997]. Columns are typically long tubes made of metal, glass, polytetrafluoroethylene, or fused silica. Columns in portable, direct-reading GCs are of two kinds: packed and wall-coated. A packed column contains a granular material used as a solid support, which is coated with a chemical chosen for its ability to interact with the components of the sample. This chosen chemical is referred to as the *stationary phase*. Packed columns are generally from 4 cm or 5 cm to 1 meter (m) or more in length and have external diameters on the order of 0.3 cm (⅛ inch). A wall-coated column tends to be longer (5 cm to 3 m or more) and narrower (internal diameter (i.d.) from 0.1 millimeter (mm) to 1 mm) than packed columns. In a wall-coated column, there is no granular solid support for the stationary phase. It is, as the name implies, coated directly on the inner walls of the column. The long, thinner columns (i.d. < 0.5 mm) are sometimes referred to as capillary columns.

The sample is carried through the column by an inert (relative to the sample) carrier gas, which, depending on the direct-reading GC, may be helium, hydrogen, nitrogen, argon, carbon dioxide, or air. The separation is governed by the degree of interaction of the sample with the stationary phase and the properties of the carrier gas. All components of a mixture spend the same amount of time in the carrier gas, so their different elution times are a function of the time partitioning between the stationary phase and the gas phase. The elapsed time from injection until the detector senses a component of a mixture is that component's retention time. The retention time is a function of the physical properties of a component in a sample, whereas the size of the peak is a function of the amount. The degree of separation of two components and their relative retention times depends, in part, on the temperature at which the system operates: the higher the temperature, the shorter the retention times. Some portable GCs operate only at ambient temperatures; others are capable of heating the column.

As each of the component parts of a mixture elutes from the column, it goes into the detector. Portable GC detectors include flame ionization, photoionization, electron capture, ultraviolet, flame photometric, thermal conductivity, nitrogen-phosphorus, and argon ionization (which have already been addressed earlier in Part II).

Because of their separation capabilities, GCs offer excellent selectivity combined with low LOD. The LOD is primarily a property of the individual detectors and are given in the detector discussions, but portable GCs generally have an LOD at sub-ppm levels. Some limitations associated with portable GCs include size, cost, and the need for more user knowledge of the technique.

Mass Spectrometers

Mass spectrometers determine the mass of molecular fragments. Specifically, a mass spectrometer determines the masses of individual fragments that have been converted into ions. A mass spectrometer determines mass by measuring the mass-to-charge ratio of ions formed from the molecule(s). After the ions are formed, they are separated in the mass analyzer according to their mass-to-charge ratio and collected by a detector wherein the ion flux is converted into an electrical signal proportional to the ion flux [ASMS 1998].

Separation by, for example, gas chromatography of the components in a mixture prior to mass spectral analysis usually provides for unambiguous identification of mixture components. Mass spectrometry is the only technique currently available that will provide for such identification of compounds in the field.

Mass spectrometers are currently used only for area samples because of their size and power requirements. They are also limited to analysis of volatile organic compounds.

Summary

Many different types of monitors are available for direct-reading analysis of gases and vapors. They operate on numerous principles of detection and vary in performance characteristics, such as linear range, specificity, sensitivity, LOM, and LOD. They are also available in many different sized packages that cover the range from personal monitoring to area monitoring. Their analysis capabilities can range from basic concentration range finding to analyses that are comparable to laboratory-based instruments and methods.

Part III. Suggested Components in Monitor Evaluation

Part III describes in detail the suggested requirements and tests that might be used to evaluate a direct-reading monitor. It divides the requirements and tests into physical, operational, and performance categories, and discusses how the results are to be interpreted. The Physical Characteristics and Operational Characteristics sections provide criteria for assessing monitor documentation and physical and operational characteristics. The third section, Performance Characteristics, addresses the testing of the monitor performance and the associated evaluation criteria. The fourth and final section, Evaluation and Documentation Reports, addresses report formats and availability.

For specific tests, the required number of repetitions and suggestions on how best to conduct the test are included. These tests should be conducted with the specific analyte(s) for which the monitor was designed. If this is not possible, then chemically appropriate surrogate analytes may be used; however, evaluation documentation should make note of this fact.

If a performance or evaluation standard developed by the ISA, American National Standards Institute (ANSI), or ASTM International (ASTM) applies to the monitor under evaluation and is more stringent than these *Components*, the more stringent criteria defined under that standard should be used to the evaluate the monitor. If the suggested requirements included in these *Components* are more stringent, then the monitor testing should address both the standard (ISA/ANSI/ASTM) and the applicable section(s) of the *Components*.

The tests described here are indicative of intramonitor variability. If these tests are per-

formed on more than one monitor of the same type, estimates of intermonitor variability can be computed (see Appendix C). The variability provides a more realistic estimate of how the user may expect the monitor to perform. A minimum of at least three similar monitors, of a given type, should be used for an evaluation.

Physical Characteristics—Suggested Documentation

Physical characteristics document such properties as the size, shape, weight, and detection method of the monitor. Also included are instrumentation documentation for operation, maintenance, and training. The following suggested requirements given in the manufacturer's documentation help the end user in the selection of an appropriate monitor to analyze certain workplace contaminants.

Documentation

Each monitor manufacturer should provide documentation on the operation, maintenance, and theory of operation for the specified monitor. The documentation may be either hard copy or electronic. It should provide an easy means to find the operating instructions for the monitor. If the monitor readout reports concentrations in ppm or related units, information should be included on the correction of these values to standard temperature and pressure conditions. If the monitor is designed to be operated by a technician, then the operating instructions should be clearly written, giving common problems in lay terms. If the monitor is designed to allow user maintenance, the maintenance procedures should be clearly specified. Any necessary parts for maintenance should be listed.

Descriptive Information

In addition to the user manual, additional information should be available on the principle of monitor operation. This information may be provided either in the operating manual for the monitor or in supplemental information. It should include a basic description of the underlying physics of detection, discussion of potential generic interferences, and applicability of the monitor. If there is need for operator sophistication beyond the level of technician, this should also be discussed. If there are Federal, State, or local regulations on the chemicals that the monitor detects, these should be discussed in terms of the applicability of the monitor.

Physical Information

To facilitate timely repair of malfunctioning monitors, the instruction manual should contain a circuit diagram if there are user serviceable parts in the monitor (e.g., fuses, etc.). At a minimum, the manual should contain a trouble-shooting guide to help the user diagnose problems and obtain the appropriate service. If user replaceable parts are available, a source or sources should be included in the manual. If possible, the manual or other documentation should state the availability of the parts. If the manufacturer has a planned obsolescence policy, it also should be stated, e.g., parts will be available for a specified number of years after the end of production of the current model. The manual should also state the manufacturer's policy regarding technical support of the monitor, its cost if available, response time, and the way that it is supplied.

Portability

Portability of a monitor can be divided into four classifications. These include the following:

- Personal: weighing in the range of 500 g or less, typically wearable by an individual.
- Portable: easily carried by one person, in the weight range of 12 kilogram (kg) or less.
- Transportable: able to be carried by two people or moveable by one person with a cart, in the weight range of 13 kg to 25 kg.
- Stationary: intended as a fixed monitor or weighing more than 25 kg.

These are guideline weights only, but serve to classify the monitors. A more accurate task specific approximation can be made using the NIOSH lifting equation [Waters et al. 1993].

Design

Design refers to the flexibility of the construction of a given monitor. For example, some electrochemical monitors are equipped to accept interchangeable electrochemical cells. These monitors typically sample by diffusion and are personal monitors. Some large units, such as portable GCs, can be fitted with interchangeable detectors. Conversely, many available monitors are fixed for a single analyte. Carbon monoxide, for example, may be monitored by a potentiometric analyzer that has an interchangeable electrochemical sensor or by a GC that is fitted with a thermal conductivity detector or an infrared detector constructed using filters to provide CO specificity. The description of the monitor design should be included in the documentation.

Safety

The manufacturer should provide instructions for the safe operation of the monitor. These should include any specific warnings about procedures or situations that may be hazardous to the operator or others in the general vicinity of an operating monitor. Safety precautions regarding calibration should be explicitly stated. The safety of the monitor in hazardous atmospheres should be stated, as well. If the monitor has been approved for use in flammable atmospheres, this should be indicated on the monitor and appropriate certifications should be provided. A common requirement for safety is compliance with the National Electrical Code definition of an intrinsically safe circuit in which any spark or thermal effect is incapable of causing ignition of a mixture of flammable or combustible material in air under prescribed test conditions [NFPA 2008].

Operational Characteristics—Suggested Documentation

Operational characteristics describe properties, such as the ease of use, maintenance, calibration, and the results of ruggedness testing. The following suggested requirements should be documented by the manufacturer.

Calibration

The manual should state whether the monitor is calibrated by a single point or multipoint calibration procedure. It should state all necessary components (mechanical and chemical) needed for monitor calibration and the concentration range of calibration. The linear range of the calibration should be stated along with sensitivity (slope of the calibration curve) if applicable. The effects of temperature and pressure on calibration should be discussed along with any required correction factors for these variables. The manual should recommend procedures, if any, for quality assurance checks on calibration.

Ease of Use

The sophistication of the operator should be indicated in the operator's manual and any necessary operator training should be specified. For transportation into field situations, any special Department of Transportation/Nuclear Regulatory Commission/State Regulations regarding the transportation of the monitor should be addressed in the manual. These should also include international shipping requirements where applicable. The warm-up time for optimal operation of the monitor should be stated.

Alarm

The type of alarm, if applicable, should be stated along with its alarm characteristics and specifications. These should include operational characteristics for either audible or visual alarms, such as noise level, intermittent/constant alarm, alarm duration, audible/visual alarm level, alarm set points and differentiation between set points, reset procedures, alarm reproducibility and accuracy, and alarm recovery when challenge concentration is removed. Any manufacturer recommended checks on alarm performance should be included as well.

Power or Battery

The power requirements for operation of the monitor should be specified in the manual (e.g., 12 volt battery, 110 volt AC). The ability of the monitor to function when there are power surges should be discussed. For battery-powered monitors, operational span on a fully charged battery should be specified along with the operational temperature range. With rechargeable batteries, recharging time and ability to recharge while running on line power should be indicated. The life span of the batteries should also be specified. The monitor should have a defined indicator to show battery charge condition and readiness for operation. The accuracy of such an indicator should be estimated.

Readout

For monitors equipped with a readout (digital or analog), the readout should provide consistent and accurate (see Appendix A) information about the response of the monitor to the challenge atmosphere. The readout should provide an easy-to-read indication of the concentration of the compound of interest. The increments on the readout scale should be easily viewed and interpreted. Nonlinear scales or indicators can be used in the monitor readout, but the instructions for interpretation should be explicit.

Data Reduction

The mechanism by which data are collected and computed should be described. The frequency of sampling should be stated and the data manipulation should be described. The data handling algorithms should be valid and appropriate for the instrumental purposes. If correction factors are required for given conditions, these should be specified and the calculations used to compute them should be verified.

Performance Characteristics

Performance characteristics, the third step in direct-reading monitor evaluation, are determinations of sensitivity, specificity, response or recovery time, and response to interferences (chemical, electromagnetic, and environmental [temperature and humidity]). Recommended means for testing these characteristics are detailed below. These tests should be conducted on monitors that are representative of production monitors intended for commercial use. If performing many of the tests, it may be possible to combine some of the tests for efficiency.

Generating an accurate atmosphere of known concentration to challenge the monitor is key to a valid evaluation. There are many different ways to generate an atmosphere [Nelson 1971, 1992] and a discussion of such goes beyond the scope of these *Components*. Often a second method of measurement must be used to verify the concentration generated. To facilitate the comparison of the independent method to the monitor results, appropriate statistical tests must be used to account for error in both the reference and test methods (see Appendices A, B, and C).

Response Time

The monitor should be calibrated using the manufacturer's instructions and manufacturer-specified calibrants if needed. The monitor should be exposed to concentrations that correspond to 20%, 40%, 60%, 80% and 100% of full scale. The time that it takes the monitor to reach 90% of the specific reading should be recorded. For example, if the monitor is exposed to a concentration equivalent to 40% of full scale, then the time that the monitor took to reach 36% of full scale should be recorded (40% × 90% = 36%). After each exposure, the monitor should be exposed to clean air. The time required to return to 10% of the initial challenge level should be recorded, as well. The times should be compared to the manufacturer's specifications. Multiple replications at each level should be performed to obtain the typical range of response times that might be expected with a given monitor.

Additionally, the monitor should be exposed to ten times the exposure limit in order to check potential problems due to saturation. The determination of response time as well as the recovery time after exposure to clean air should be determined as indicated above.

For alarm-based monitors, the exposure concentrations should be 10% and 30% above the alarm set point. Also, the reading at which the alarm sounds should be adjusted to control for false negatives (see Appendix F) and the unit should be tested at that concentration. The time after exposure at which the alarm sounds should be recorded as the response time. After each exposure, the monitor should be exposed to clean air. The time required to return to non-alarm level should be recorded, as well.

Calibration, Linearity, and Drift

The monitor should be calibrated using the manufacturer's instructions and manufacturer-specified calibrants, if applicable. (Some monitors are precalibrated at the factory or have calibration procedures that are not applicable in the field environment.) The zero point of the monitor should be adjusted, if adjustable, to zero in clean air. All the manufacturer's safety instructions should be followed and care should be taken when working with toxic calibrants to avoid exposure. After initial calibration, the monitor should be allowed to sit in operating mode in clean air for at least 4 h. After that period of time, the monitor should be challenged with the calibrant gas for a period of time that exceeds the response time (see Response Time above) by at least a factor of two. Some monitors may require additional time to stabilize. The monitor should respond to the challenge, giving an indication of concentration that is within ±10% of the certified value of the calibrant. Once again, the monitor should be allowed to sample clean air for at least 30 min and, again, challenged to the

calibrant for a period of time that exceeds the response time by at least a factor of two. All monitor responses to the challenges should indicate concentrations within ±10% of the certified value of the calibrant.

The linearity of the monitor should be checked by exposing the calibrated monitor to concentrations equivalent to the points used for the calibration and allowing the monitor to stabilize. Additional concentrations beyond the calibration concentrations can be used to confirm linearity. Deviation of monitor response from the calibration concentrations should not exceed 10%. This criterion may be difficult to meet at low concentrations. Failure to meet the criterion at low concentrations should be noted.

After exposure to clean air, the drift of the monitor should be checked by exposing the monitor to known concentrations of analyte and allowing the monitor response to stabilize. The monitor reading should be within ±10% of the known concentration of the analyte. The monitor should again be exposed to clean air and the monitor response allowed to stabilize. Plot results versus time to observe trends in response to known concentrations and clean air. This study should be repeated several times on the first day and on consecutive days. Both high and low concentrations within the response range of the instrument should be used for this study. The results should be plotted over time to observe any trends of zero and/or span drift.

For alarm-based monitors, concentrations 10% above and below the alarm set points should be used for testing drift. The response of the monitor should be studied over several days to observe any potential drift.

Monitor bias and variability must be characterized in order to control false positives and negatives. In the course of any 20 exposures, only one false positive and/or one false negative should be found, though achieving this goal may depend on the exposure concentration. Appendices F and G give, respectively, a method and a software program for controlling the number of false positives and negatives for alarm systems. These depend on bias and variability of the monitor response. It is desirable that the false positive and negative rates be controlled at some reasonable rate, say, 5%.

Range

The range of the monitor is defined by the scale of the readout device. Some monitors have multiple scales, extending the range of the device. If the analyte used for testing has a legally defined exposure limit, the monitor range should be verified at 0.1, 0.5, 1.0, and 2.0 times that limit, with at least four replicates at each concentration. If there are multiple limits for a compound, such as TWA and STEL values, the lower limit should be used to calculate the lower end of the range and the higher limit should be used to calculate the higher end of the range. The monitor should respond at each concentration within ±10% of the challenge concentration. If the monitor range extends beyond these defined concentrations, then an additional point for each additional scale should be included. The same criterion of ±10% response applies to these additional points.

Some monitors provide a unique challenge to the concept of range. For example, combustible gas monitors are designed to respond to a percentage of the lower explosive limit, and some monitors capable of monitoring multiple analytes may have an analytical range that varies by analyte. These monitors provide unique challenges for evaluation and may require specialized concentration generation equipment to address these issues.

Environmental Effects

The monitor should be exposed to extremes of temperature and humidity, as defined by the manufacturer, as well as to intermediate temperature and humidity. The most appropriate approach for this test is based on factorial experimental design [Box et al. 1978]. Under each of these conditions, the monitor should be challenged with concentrations of 0.1, 0.5, 1.0, and 2.0 times the legal exposure limit or other appropriate limits (e.g., LEL for a combustible gas monitor) of the analyte under study. If no exposure limit exists for the analyte, then three concentrations should be used that are representative of low, intermediate, and high concentrations. The monitor response should be within ±10% of the known challenge concentration. If the manufacturer has supplied a correction factor for temperature and/or humidity, then this correction factor should be applied to correct the monitor reading before comparison with the challenge concentration. Statistically significant effects should be reported and appropriate correction factors provided by the manufacturer. If the monitor uses a diffusional sampling principle, the effect of face velocity (low, intermediate, and high) on the monitor performance should be evaluated [Rose and Perkins 1982]. Doing this entails the design of an exposure chamber that has a control system that maintains a consistent face velocity during an exposure experiment.

Precision

Precision should be calculated from exposure experiments at a given concentration. (See Part III, Performance Characteristics [Response Time; Calibration, Linearity, and Drift; Range; and Environmental Effects], listed above.) The monitor precision at each concentration then

Components for Evaluation of Direct Reading Monitors

should be checked for statistical homogeneity and pooled with the precisions from other concentrations. (Statistical procedures are given in Appendices B and C.) The pooled estimate will give an overall monitor precision value. If the precisions cannot be pooled, then the precision should be reported at each concentration. Incorporation of intermonitor variability into the precision calculations is addressed in Appendix C.

Bias

Bias should be determined for all the individual exposure experiments. (See experiments described in Part III, Performance Characteristics, [Response Time; Calibration, Linearity, and Drift; Range; and Environmental Effects], listed above). Biases should be checked for homogeneity and pooled if possible (see Appendices B and C for details). If the bias is not constant over the evaluation range, then the bias may be reported for the individual concentrations used in the evaluation, although it may be possible to pool smaller groups of biases (see Appendices B and C). If the magnitude of the estimated bias exceeds 10%, then the measurement should be bias-corrected (see Appendix A). Incorporation of intermonitor variability into the bias calculations is addressed in Appendix C. For alarm-based monitors, see Appendices F and G.

Accuracy

Calculation of accuracy is based on the precision and bias measured for a given monitor. If bias and/or precision cannot be pooled statistically, then the accuracy can be expressed for individual concentrations. Details for these calculations are given in Appendix A, and procedures for pooling precision and bias are discussed in Appendices B and C. An alternative test of acceptable accuracy for either unbiased or bias-corrected methods is that the probability that a measurement is within ±10% of the true concentration should be at least 56.7%. See Appendix E for the details of this test. For alarm-based monitors, see Appendices F and G.

Limit of Measurement

The scale used for the readout of the monitor can be used to define the LOM of the monitor. This limit should be verified by exposing the calibrated and operating monitor to an atmosphere that contains a concentration of analyte that is 10 times above the expected LOM of the monitor (typically equivalent to the smallest division or gradient on the monitor readout). The monitor should register concentration on its readout within ±10%. This exposure should be conducted 10 times. If the monitor does not meet this criterion eight out of ten times, the LOM should be incrementally raised until the monitor can meet the performance criterion.

Using the American Chemical Society [ACS Subcommittee on Environmental Improvement 1980] definitions of LOD and limit of quantitation (LOQ) is an alternative to quantitation of the LOM. Under this approach, the signal to noise level for the instrument is measured in clean air. The LOD is defined as three times the signal to noise level. The LOQ (which is related to the LOM) should be ten times the signal to noise level. These calculated levels should be verified by exposure to concentrations equivalent to these levels. Whereas at the LOD, the monitor should detect the analyte 50% of the time, for blank samples, detection should occur about 1% of the time. At the LOQ, the monitor should quantitate the analyte with defined precision and accuracy.

Noise is the inherent level of instrumental response attributable to electronic and other systems within the monitor. Signal from background sources may also be considered noise. In any monitor, it is ideal to minimize background and monitor noise and to maximize the signal to noise ratio.

Environmental Interferences

The effect of documented interferences on the operation of the monitor should be tested. The measured magnitude of the interference as indicated by the monitor readout should agree with manufacturer specifications and documentation. If the monitor is to be used in a specific environment, then the effect of that environment on monitor performance should be checked.

Electromagnetic Interference

Feldman [1993] reported that many industrial hygiene sampling instruments were subject to electromagnetic susceptibility problems, resulting in malfunction or error, such as false alarm [Cook and Huggins 1984]. In most cases, such problems are mitigated by conformance to the ANSI/ISA [2010] requirement: "Gas-detection apparatus, their components, and remote detector heads must be constructed to be resistant to, or protected against, electromagnetic interference. Testing shall be verified and documented in accordance with EN 50270 by an ISO/IEC 17025 accredited test laboratory." However, the scope of EN 50270 [CENELEC 2006] includes the note: "In special cases, situations will arise where the level of disturbances may exceed the levels specified in this standard, e.g. where an apparatus is installed in proximity to industrial, scientific or medical (ISM) equipment as specified in EN 55011 or where a hand-held transmitter is used in close proximity to an apparatus. In these instances special mitigation measures may have to be employed." If a monitor is to be used in close proximity to possible sources of electromagnetic interference, such as hand-held transmitters, then the effect of that electromagnetic environment on monitor performance should be checked.

Drop and Vibration

Drop and vibration tests of the monitor should be conducted according to the specifications outlined in the ANSI/ISA standard (toxic gas detectors) [ANSI/ISA 2010]. Failure of the monitor to operate after either the drop or the vibration tests indicates failure of the monitor in the evaluation process. No further testing should be done.

Remote Sampling

Remote sampling describes the ability of a monitor to obtain a sample from other than the monitor's immediate surroundings. It is useful for obtaining samples from hazardous environments, but does raise issues of increased response time and of sample loss through sample transport lines [Dowker and Hardwick 2008; RAE Systems 2006, 2010]. At times, inert sample lines can be used to reduce or eliminate wall adsorption or loss, or it may be possible to heat sample transfer lines to eliminate sample loss from condensation.

To investigate the possible loss of analyte in sampling lines, the monitor should be calibrated according to the manufacturer's instructions. The monitor should sample known concentrations of analyte directly without the remote sampling line and, then, again with the sampling line included. Any difference between these two sets of values can be represented as potential sample loss due to transfer line. Sample losses due to transfer line should not be greater than 10% of the initial concentrations. Estimated sample loss induces negative bias in measurements that should be included in accuracy calculations. (See Appendix A.)

Detector Life

As a minimum, the manufacturer should provide an estimate of the expected detector life in terms of operational hours before the monitor performance falls below a stated accuracy criterion. While experimental verification of this time period may not be practical in the evaluation of the monitor, recording operation and maintenance in a monitor logbook will verify detector lifetime estimates. To observe any effects on monitor accuracy in both new and older detectors, evaluation of monitor performance is recommended, per experiments described in Part III, Performance Characteristics. This evaluation will help in establishing the expected detector life discussed above.

Step Change Response and Recovery

This test evaluates the response and recovery of the monitor during intermittent exposures to high concentrations. The test should be conducted in accordance with the procedures outlined in the ANSI/ISA standard (toxic gas detectors) [ANSI/ISA 2010].

Supply Voltage Variation

This test evaluates the performance of the monitor when supply voltages are varied by 85% to 110% of nominal voltage required for operation. This test should be performed in accordance with the procedures outlined in the ANSI/ISA standard (toxic gas detectors) [ANSI/ISA 2010].

Long-Term Stability

Since long-term testing (3 months to 12 months) may not be feasible, recording monitor performance, quality assurance, and calibration parameters in a logbook during monitor use provides an indication of the long-term performance of the monitor. Periodically, the logbook should be checked for any major changes in calibration or monitor sensitivity. Plotting data in the logbook can show trends in monitor performance.

Monitor Uncertainty

The monitor uncertainty [ISO 1993] can be related to terms of bias and precision. The standard deviation of a measurement is the standard uncertainty of that measurement. Accuracy as defined in this document relates to the expanded uncertainty of the instrument with a coverage factor of 95%. (See Part III, Performance Characteristics, Accuracy and Appendix D.) Incorporation of intermonitor variability into the precision calculations is addressed in Appendix C.

Quality System Requirements

The test facility that undertakes the evaluation of monitors should meet the requirements of the International Organization for Standardization/International Electrotechnical Commission standard ISO/IEC 17025:2005 General Requirements for the Competence of Testing and Calibration Laboratories [ISO 2005]. Although this document does not require ISO 17025 certification, it recommends compliance with the principles of that standard. Compliance with this standard means that the test facility has implemented a quality system and has the staff and facilities competently perform the evaluations required. Test facilities that meet ISO/IEC 17025:2005 requirements also comply with ISO 9001 or 9002 requirements, depending upon the activities they undertake.

The evaluation of monitors is based upon the assumption that the manufacture of monitors is under statistical quality control and free of manufacturing changes and defects that affect monitor performance. Based on an ISO standard [ISO 2005], a system must be established for reporting instances of monitor failure or shortcomings from the field. Information on complaints, the occurrence and modes of failure, and customer needs and expectations should be investigated and appropriate corrective actions undertaken. While this standard [ISO 2005] does not specify an overall monitor manufacturing quality system requirement, many monitor manufacturers meet ISO 9000 series quality system requirements. ISO 9001, ISO 9002, and ISO 9004 quality system requirements meet the reporting, investigation, and corrective action of customer complaints requirements of this evaluation.

Reliability

Reliability of a monitor means a statistical comparison of performance among multiple, similar monitors (intermonitor reliability); the performance of the same monitor to repeated, identical challenges (intramonitor reliability); or the response of a monitor on successive days (day-to-day reliability). Appendices B and C address the treatment of the intra- and inter-monitor variability.

Field Evaluation

It is beneficial to compare measurements by the monitor unit with measurements from an independent method in a field study. Bias relative to the independent method and precision of the unit can be determined from the field data. (See Appendix B, Section 2.4.)

Monitor Results

Monitor results from the exposure experiments described above should be normalized to 100% $\left(C_{\text{measured}}/C_{\text{true}}\right)$ for a given concentration. C_{true} can be replaced by C_{ref}, if the mean of an independent method is used. See Appendix D for evaluation of monitor uncertainty for each of these cases.

- **Day-to-day:** The normalized monitor results should be graphed versus day. The expected result is a horizontal line or piecewise horizontal line (different line for each day) between 90% and 110%. The ordinate of the horizontal line estimates the bias, which is not to exceed 10%. Precision over days should be estimated, which correspond to the variability about the lines. (See Section 4 of Appendix B.)

- **Intraday:** The normalized monitor results should be graphed versus true concentration or estimated true concentration. The expected result is a horizontal line between 90% and 110%. Intraday precision should be estimated, which corresponds to the variability about the line. (See Section 4 of Appendix B.)

- **Intermonitor:** Precision of the normalized results from multiple monitors (if available) should be evaluated. (See Appendix C.)

- **Intramonitor:** Precision of the normalized results from a single monitor should be evaluated. (See Appendix B.)

Monitor Evaluation Data Reduction

Monitor evaluation data reduction addresses data handling and statistical interpretation of the data obtained from the testing. This section focuses on necessary evaluation mathematics and statistics, but the more complex mathematical treatments appear in the appendices. A summary of the experiments suggested above is included in Table 1. For readout-equipped monitors, the data should be accessible in a readable form, either electronically or visually. The information may be a direct representation of the sampled concentration or a relative reading to a calibration standard(s). In addition, Table 2 provides a summary of monitor physical and ease-of-use characteristics, based on evaluation by a panel of experts as described in NIOSH [2012].

Table 1. Direct-reading monitor evaluation experiment summary

Performance characteristic	Experiment	Criterion
Response time	Expose to 4 concentrations (20%, 40%, 60%, 80%, and 100% of full scale); use multiple replicates to provide better estimate of range; return monitors to clean air between each exposure.	Record time for meter to reach 90% of concentration value; record time to return to 10% of baseline; verify manufacturer specifications.
Calibration	Zero and calibrate; clean air for 4 h; exposures at each concentration used for calibration for at least 2 times the response time; allow 30 min sampling clean air between exposures; perform 2 replicates.	Results should be between 90% to 110% of expected values.
Linearity	Exposures at each concentration point on calibration curve and allow response to stabilize; additional points can be added.	Results should be between 90% to 110% of expected values.
Drift	Check monitor response to clean air and to known analyte concentrations; additional exposures should be performed over several days.	The monitor reading should be within ±10% of the known concentrations of the test analyte; plot results versus time to observe trends in drift for both zero and test analyte concentrations.
Range	Exposure at 0.1, 0.5, 1.0, and 2.0 times the exposure limit for a given chemical; if monitor scale extends beyond these concentrations, additional points should be added to test range; a minimum of 4 replicates at each point should be included.	Monitor should respond within ±10% of each of these values.
Environmental effects	Expose at 4 concentrations (0.1, 0.5, 1.0, and 2.0 times the exposure limit), at temperature and relative humidity intermediate and extreme levels.	Monitor response should be within 90% to 110% of expected concentration value (temperature and relative humidity correction applied if recommended by manufacturer).
Precision	Evaluate precision; if analyte has legal limit, then the concentrations should be 0.1, 0.5, 1.0, and 2.0 times the limit.	Calculate precision for given concentration levels from experiments above; check homogeneity and pool values if possible.

(Continued)

Table 1 (Continued). Direct-reading monitor evaluation experiment summary

Performance characteristic	Experiment	Criterion
Bias	Calculate bias values for each concentration level; check homogeneity and pool values if possible. Use same concentration values as given for "precision" if analyte has a legal limit.	Estimated bias should be no more than 10%. If this criterion is not met, measurements must be bias-corrected.
Accuracy	Combine precision and bias values as shown in Appendix A.	Accuracy should be ≤25% with 95% confidence.
Limit of measurement	Exposure at concentration that is 10 times the smallest increment on the monitor readout for 10 times.	Concentration reading should be within 10% of the exposure concentration; keep raising challenge concentration until monitor meets the above criterion.
Environmental interferences	Verify any potential interferences present in environment to be sampled. For use in harsh environments, interferences can include particles, smoke, fog, dust, gases, fuel vapors, aqueous film forming foam, household chlorine bleach, engine exhaust.	If environmental interference bias exceeds magnitude of 10%, the monitor fails.
Electromagnetic interference	For tests done in close proximity to possible sources of electromagnetic interference, the monitor's displayed values should be compared to a known concentration (chosen between the LOQ and IDLH levels) at all frequencies and closest distances likely to be used.	When tested within the tester-specified distances from, for example, a 5 watt UHF/VHF source at specified frequencies, the monitor's displayed values should differ by no more than 10% from the true value, whether the electromagnetic interference is present or absent.
Drop and vibration	Drop and vibration tests per the ANSI/ISA standard (toxic gas detectors) [ANSI/ISA 2010].	Pass tests.
Remote sampling	Sample concentrations with and without remote sampling line.	Concentration data with and without remote sampling line should agree within 10%.
Detector life	Determine how long the detector lasts.	Obtain manufacturer data.
Step change response and recovery	Test in accordance with ANSI/ISA standard (toxic gas detectors) [ANSI/ISA 2010].	Pass test.

(Continued)

Table 1 (Continued). Direct-reading monitor evaluation experiment summary

Performance characteristic	Experiment	Criterion
Supply voltage variation	Test in accordance with ANSI/ISA standard (toxic gas detectors) [ANSI/ISA 2010].	Pass test.
Long-term stability	Determine if there is drift in instrumental response to known amount of analyte.	Plot repeat measurements over a long period of time.
Monitor uncertainty	Provide uncertainty.	Express uncertainty either as a percentage or as an interval (Appendix D).
Quality system requirements	Follow guidance in ISO 17025–2005 [ISO 2005], ISO 9000 series quality systems.	
Reliability	Combine monitor exposure data and normalize by concentration; plot data versus time for each concentration; look at day-to-day, inter- and intra-monitor variability separately.	Variance components should be estimated. Plots of normalized results should be piece-wise horizontal lines between 90% and 100%. (See Appendix B.)
Field evaluation	Compare measurements by the monitor unit with measurements from an independent method in a field study.	Bias relative to the independent method, and precision of the unit should be determined. (See Appendix B, Section 2.4.)

Table 2. Assessment of physical characteristics and ease of use—may be best carried out via an evaluation panel

Characteristic	Goal of evaluation
Documentation	Operation, maintenance, and theory of operation for monitor.
Transport mode	How user will carry it; relates to any protective clothing worn.
Ease of decontamination	Methods used should be applicable to the type of exposure anticipated.
Operational controls	Turning on/off; zeroing; internal calibration check; gas selection; reading result.
Alarms	Audible/vibration systems; visual alarms; low battery indicator.
Visual display	Digital readout in consistent units; evaluate brightness and size of display.
Maintenance	Replaceable power supply, replaceable detector.

Data Requirements

Selection of a specific direct-reading monitor for monitoring is dependent, in part, on the user's data requirements relative to the monitoring results, or data. That is to say, how will the data be used? How accurate must the data be? Data quality objectives are defined by a number of parameters including, but not limited to, the following.

- **Purpose of the monitoring:** The reason for monitoring should be known—whether it is exposure assessment driven, and/or compliance driven, and/or process control driven, and/or environmental quality driven.

- **Proposed use of the data:** Determination of how the data will be used is necessary—whether for a TWA, STEL or ceil limit, or whether as an alarm warning instead of a quantity. Are qualitative or quantitative results required?

- **Analytical limitations of the monitor:** Monitor analytical limitations should be known—whether it can deliver the needed data, such as sufficient accuracy, precision, and LOD or LOQ.

- **Conditions of sampling:** Conditions of sampling may dictate the kind of monitoring and, in turn, the kind of data obtainable. For example, if the concentration is rapidly changing, a monitor that captures a single grab sample is inappropriate since a snapshot data response will not give an accurate representation of what is occurring, although a series of data snapshots might better describe the exposure. A continuous monitor would provide a better indication of exposure profile.

- **Any legal requirements on the data:** If data are to demonstrate compliance with an OSHA PEL or other regulatory exposure limits, certain statistical requirements may be placed on them, which may, in turn, influence and/or limit the choice of the monitoring instrument used. Typically, if the monitor meets the accuracy requirements as defined in this document, the monitor will meet most compliance requirements.

Evaluation and Documentation Reports

Evaluation and documentation reporting is the final step in the evaluation of direct-reading monitors. A suggested format for reporting results is given below. The intent is to make monitor testing and reporting as consistent as possible.

The report should document all the experimental work performed and monitored. A summary of the data should be included in the report. The report should include clear, concise statements on the performance of the monitor for each of the evaluation steps.

The report should contain the following sections.

- **Introduction:** Description of the monitor evaluated.

- **Experimental:** Description of generation system and experimental procedures.

- **Results:** Description of the experimental results.

- **Discussion and conclusions:** Discussion of the monitor results, their interpretation, and conclusions made about the monitor performance.

The report should be published in an accessible location. Publication in peer-reviewed literature is the most desirable means and also helps to ensure impartiality in the data presented and its interpretation.

Components for Evaluation of Direct Reading Monitors

Relevant Standards

ANSI/ISA-12.13.01-2002: Performance Requirements for Combustible Gas Detectors. Research Triangle Park, NC: American National Standards Institute, ISA—The Instrumentation, Systems, and Automation Society. [http://www.isa.org/Template.cfm?Section=Standards8&Template=/Ecommerce/ProductDisplay.cfm&ProductID=6740]

ANSI/ISA-92.00.01-2010: Performance Requirements for Toxic Gas Detectors. Research Triangle Park, NC: American National Standards Institute, International Society of Automation. [http://www.isa.org/Template.cfm?Section=Standards8&Template=/Ecommerce/ProductDisplay.cfm&ProductID=11549]

ANSI/ISA-92.04.01, Part I-2007: Performance Requirements for Instruments Used To Detect Oxygen-Deficient/Oxygen-Enriched Atmospheres. Research Triangle Park, NC: American National Standards Institute, ISA—The Instrumentation, Systems, and Automation Society. [http://www.isa.org/Template.cfm?Section=Standards8&Template=/Ecommerce/ProductDisplay.cfm&ProductID=10006]

EN 51270:2006. Electromagnetic compatibility - Electrical apparatus for the detection and measurement of combustible gases, toxic gases or oxygen. Brussels, Belgium: European Committee for Electrotechnical Standardization. [http://shop.bsigroup.com/en/ProductDetail/?pid=000000000030145393]

References

ACGIH [1992]. 1992–1993 Threshold limit values for chemical substances and physical agents & biological exposure indices. Cincinnati, OH: American Conference of Governmental Industrial Hygienists.

ACS (American Chemical Society) Subcommittee on Environmental Improvement [1980]. Guidelines for data acquisition and data quality evaluation in environmental chemistry. Anal Chem 52:2242–2249.

Anderson C, Gunderson E, Coulson D [1981]. Sampling and analytical methodology for workplace chemical hazards. In: Choudhary G, ed. Chemical hazards in the workplace: Measurement and control. ACS symposium series, 149. Washington, DC: American Chemical Society, pp. 3–19.

ANSI/ISA [2010]. ANSI/ISA-92.00.01-2010: Performance requirements for toxic gas detectors. Research Triangle Park, NC: American National Standards Institute, International Society of Automation.

ASMS [1998]. What is mass spectrometry? 3rd ed. Santa Fe, NM: American Society for Mass Spectrometry.

Box GEP, Hunter WG, Hunter JS [1978]. Statistics for experimenters: An introduction to design, data analysis, and model building. New York: John Wiley & Sons, pp. 306–432.

Busch K, Taylor D [1981]. Statistical protocol for the NIOSH validation tests. In: Choudhary G, ed. Chemical hazards in the workplace: Measurement and control. ACS symposium series, 149. Washington, DC: American Chemical Society, pp. 503–517.

CEN (European Committee for Standardization) [1994]. EN 482 Workplace atmospheres—general requirements for the performance of procedures for the measurement of chemical agents. Brussels, Belgium: European Committee for Standardization.

CENELEC (European Committee for Electrotechnical Standardization) [2006]. EN 51270: Electromagnetic compatibility - Electrical apparatus for the detection and measurement of combustible gases, toxic gases or oxygen. Brussels, Belgium: European Committee for Electrotechnical Standardization.

CFR. Code of Federal regulations. Washington, DC: U.S. Government Printing Office, Office of the Federal Register.

Cook CF, Huggins PA [1984]. Effect of radio frequency interference on common industrial hygiene monitoring instruments. Am Ind Hyg Assoc J 45(11):740–744.

David DJ [1974]. Gas chromatographic detectors. New York: Wiley Interscience.

Dean WFH, Stockdale MJ [2006]. Ionization devices. U.S. Patent 7,046,012. Washington, DC: U.S. Patent and Trademark Office. [http://patft.uspto.gov/netacgi/nph-Parser?Sect1=PTO1&Sect2=HITOFF&d=PALL&p=1&u=%2Fnetahtml%2FPTO%2Fsrchnum.htm&r=1&f=G&l=50&s1=7046012.PN.&OS=PN/7046012&RS=PN/7046012]

Dowker KP, Hardwick K [2008]. Effect of tubing type on gas detector sampling systems. Buxton, Derbyshire, UK: Health and Safety Executive, Health and Safety Laboratory, Research Report RR635. [http://www.hse.gov.uk/research/rrpdf/rr635.pdf]

Feldman RF [1993]. Degraded instrument performance due to radio interference: Criteria and standards. Appl Occup Environ Hyg 8(4):351–355.

Fowler WK, Bradley EL Jr. [1990]. Laboratory control of analytical method inaccuracy

Components for Evaluation of Direct Reading Monitors

with respect to a given inaccuracy standard. Am Ind Hyg Assoc J *51*(3):127–131.

Gentry SJ [1993]. Instrument performance and standards. Appl Occup Environ Hyg *8*(4):260–266.

Hill HH Jr., Baim MA [1982]. Ambient pressure ionization detectors for gas chromatography. Part I: flame and photoionization detectors. Trends Anal Chem *1*(9): 206–210.

HNU Systems [1986]. Instruction manual: Gas chromatograph, model 301. Newton Highlands, MA: HNU Systems, Inc.

Hodgeson JA [1974]. A review of chemiluminescent techniques for air pollution monitoring. Toxicol Environ Chem Rev *11*:81–90.

Holm T [1997]. Mechanism of the flame ionization detector II. Isotope effects and heteroatom effects. J Chromatogr A *782*(1):81–86.

Ingle JD Jr., Crouch SR [1988]. Spectrochemical analysis. Englewood Cliffs, NJ: Prentice-Hall.

ISA [2003]. The automation, systems, and instrumentation dictionary. 4th ed. Research Triangle Park, NC: ISA—The Instrumentation, Systems, and Automation Society.

ISO [1993]. Guide to the expression of uncertainty in measurement. Geneva, Switzerland: International Organization for Standardization.

ISO [2005]. ISO/IEC 17025 General requirements for the competence of testing and calibration laboratories. Geneva, Switzerland: International Organization for Standardization.

McNair HM, Miller JM [1997]. Basic gas chromatography. New York: John Wiley & Sons.

Nelson GO [1971]. Controlled test atmospheres: Principles and techniques. Ann Arbor, MI: Ann Arbor Science.

Nelson GO [1992]. Gas mixtures: Preparation and control. Ann Arbor, MI: Lewis Publishers.

NFPA [2008]. NFPA 70: national electrical code, 2008 ed. Quincy, Massachusetts: National Fire Protection Association, section 504.2.

NIOSH [1980]. Development and validation of methods for sampling and analysis of workplace toxic substances. By Gunderson EC, Anderson CC. Cincinnati, OH: U.S. Department of Health and Human Services, Centers for Disease Control, National Institute for Occupational Safety and Health, DHHS (NIOSH) Publication No. 80–133, NTIS No. PB89–182042.

NIOSH [1992]. NIOSH recommendations for occupational safety and health: Compendium of policy documents and statements. Cincinnati, OH: U.S. Department of Health and Human Services, Centers for Disease Control, National Institute for Occupational Safety and Health, DHHS (NIOSH) Publication No. 92–100. [http://www.cdc.gov/niosh/docs/92-100/pdfs/92-100.pdf]

NIOSH [1995]. Guidelines for air sampling and analytical method development and evaluation. By Kennedy ER, Fischbach TJ, Song R, Eller PM, Shulman SA. Cincinnati, OH: U.S. Department of Health and Human Services, Centers for Disease Control and Prevention, National Institute for Occupational Safety and Health, DHHS (NIOSH) Publication No. 95–117. [http://www.cdc.gov/niosh/docs/95-117/pdfs/95-117.pdf]

NIOSH [2004]. NIOSH respirator selection logic. Cincinnati, OH: U.S. Department of Health and Human Services, Centers for Disease Control and Prevention, National Institute for Occupational Safety and Health, DHHS (NIOSH) Publication No. 2005–100.

[http://www.cdc.gov/niosh/docs/2005-100/pdfs/2005-100.pdf]

NIOSH [2012]. Addendum to components for evaluation of direct-reading monitors for gases and vapors: hazard detection in first responder environments. Cincinnati, OH: U.S. Department of Health and Human Services, Centers for Disease Control and Prevention, National Institute for Occupational Safety and Health, DHHS (NIOSH) Publication No. 2012–163. [http://www.cdc.gov/niosh/docs/2012-163/]

RAE Systems [2006]. Technical note TN-140: the effect of extension tubing volume and vacuum on sample flow for RAE Systems instruments [http://www.raesystems.com/sites/default/files/downloads/FeedsEnclosure-TN-140_Tubing_Volume.pdf]. Date accessed: December 2011.

RAE Systems [2010]. Application note AP-100: RAE Systems PID training outline [http://v2010.raesystems.com/~raedocs/App_Tech_Notes/App_Notes/AP-000_PID_Training_Outline.pdf]. Date accessed: December 2011.

Rose VE, Perkins JL [1982]. Passive dosimetry—state of the art review. Am Ind Hyg Assoc J 43(8):608–621.

Scott RPW [2003]. Book 2, Chrom-Ed book series: gas chromatography [http://www.chromatography-online.org/3/contents.html]. Date accessed: December 2011.

Shotwell HP, Caporossi JC, McCollum RC, Mellor JF [1979]. A validation procedure for air sampling-analysis systems. Am Ind Hyg Assoc J 40(8):737–742.

Skoog DA, West DM, Holler FJ [1988]. Fundamentals of analytical chemistry. 5th ed. New York: Saunders College Publishing.

Stetter JR, Li J [2008]. Amperometric gas sensors—a review. Chem Rev 108(2):352–366.

Strobel HA, Heineman WR [1989]. Chemical instrumentation: A systematic approach. 3rd ed. New York: John Wiley & Sons.

Waters TJ, Putz-Anderson V, Gary A, Fine LJ [1993]. Revised NIOSH equation for the design and evaluation of manual lifting tasks. Ergonomics 36(7):749–776.

Appendix A. Estimation of Accuracy

1. Accuracy

The accuracy of a direct-reading monitor is defined as the maximum absolute error relative to the amount being measured, such that 95% of the readings will fall in this range. In other words, the coverage probability is 95%. Thus defined, accuracy can be affected by both bias and precision (see Appendices B and C for calculations and definitions).

Precision S_r is the relative variability of measurements about the mean of the population of measurements. It is calculated by dividing σ, the standard deviation of concentration measurements from a homogeneous atmosphere of known concentration C_T, by the concentration measurement mean μ. Thus, $S_r = \sigma/\mu$. Bias B is the relative discrepancy between μ and C_T, expressed as a fraction. B is given by $B = \mu/C_T - 1$.

Another measure of precision S_{rT} is the concentration standard deviation σ relative to the true concentration C_T. Thus, $S_{rT} = \sigma/C_T$; also, $S_{rT} = S_r \times (1 + B)$.

Groupings of concentration levels with constant S_r, B, and S_{rT} must be identified. Statistical procedures for obtaining such groupings are given in Appendix B.

Under the assumption that monitor responses are normally distributed, the accuracy of a monitor, denoted by A, can be uniquely determined by its bias B and precision S_{rT} through the following equation:

$$\Phi\left(\frac{B+A}{S_{rT}}\right) - \Phi\left(\frac{B-A}{S_{rT}}\right) = 0.95 \qquad (A1)$$

[AppA01], where $\Phi(x)$ is the cumulative distribution function of a standard normal variable. (Equation (A1) is also true with B replaced by $-B$.) If S_r

is used in the evaluation, replace S_{rT} with $S_r \times (1 + B)$ in the above (A1) and the following corresponding formulas (A2, A3, A4).

Equation (A1) can be solved numerically for the accuracy A. Furthermore, A can be expressed [Krishnamoorthy and Mathew 2009] in terms of functions available as routines in various statistical packages. Alternatively, the following expansions (A2) [Bartley 2001] in the limits $B \to 0$ and $|B| \to \infty$ are recommended for calculating accuracy because of their simplicity and transparency in the dependence on bias B and S_{rT}:

$$A(B, S_{rT}) = \begin{cases} 1.96 \times \sqrt{B^2 + S_{rT}^2} & \text{if } |B| < \dfrac{S_{rT}}{1.645}, \\ |B| + 1.645 \times S_{rT} & \text{otherwise} \end{cases} \qquad (A2)$$

[AppA02].

The maximum deviation from the true accuracy value occurs around the intersector boundary with a maximum fractional error within ±1%, i.e., much smaller than would normally be expected from a simple expansion. (See Figures A–1 and A–2, where the apparent discontinuities are at the intersector boundaries. See Figure A–3 for an example of a 95% accuracy interval.)

2. Estimation

In this section confidence limits on accuracy are presented for a variety of experimental situations—known target concentration for evaluation of the test method, estimated target concentration, known relative standard deviation of the test chamber, known or estimated relative standard deviation of the independent method, and either known or unknown bias of the independent method. These situations are discussed more fully in the introduction to Appendix B.

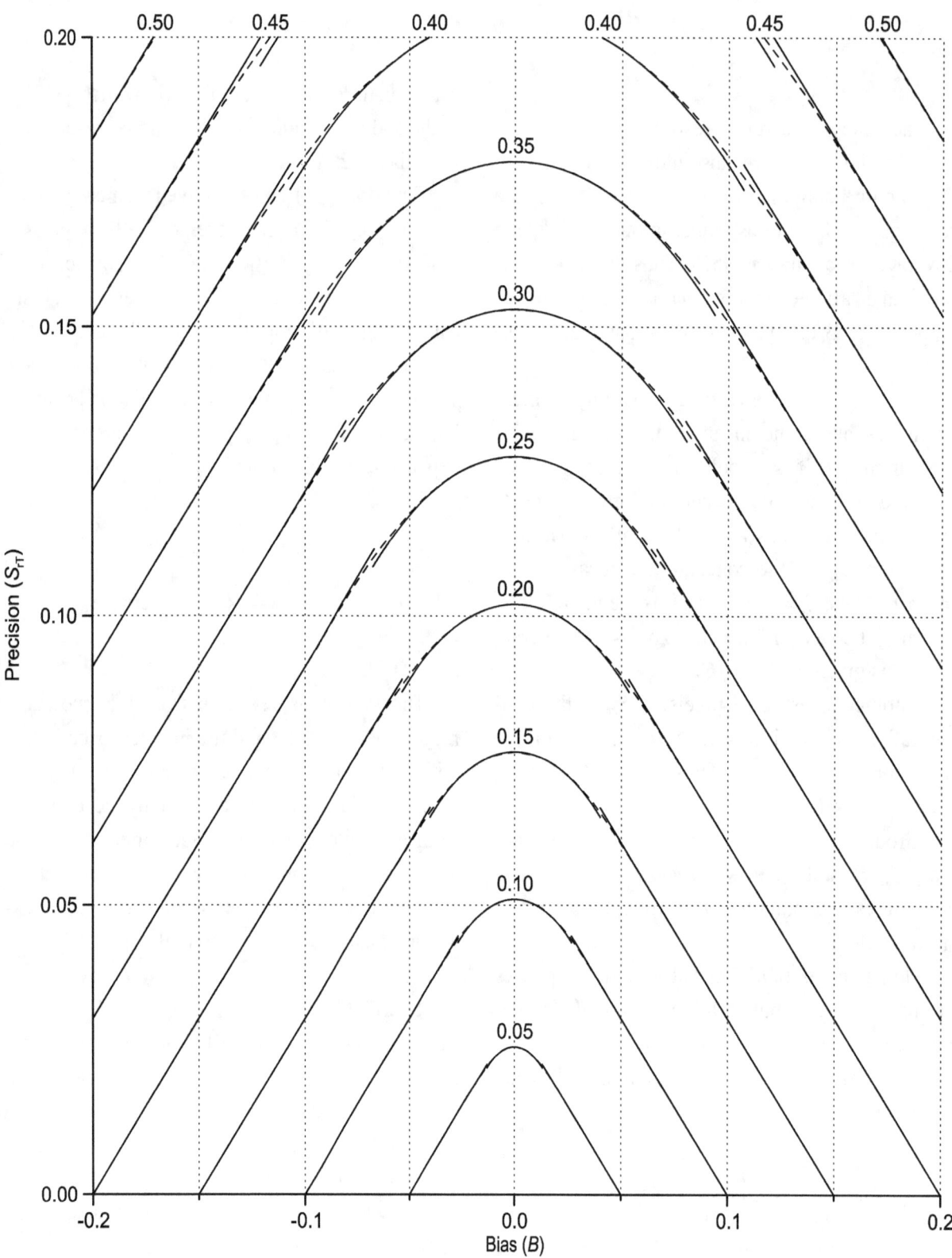

Figure A–1. Accuracy A as function of bias B and precision S_{rT} by equation (A2) (solid lines) and equation (A1) (dashed lines). Each curve shows all values of B and S_{rT} yielding A as indicated on the curve.

Components for Evaluation of Direct Reading Monitors

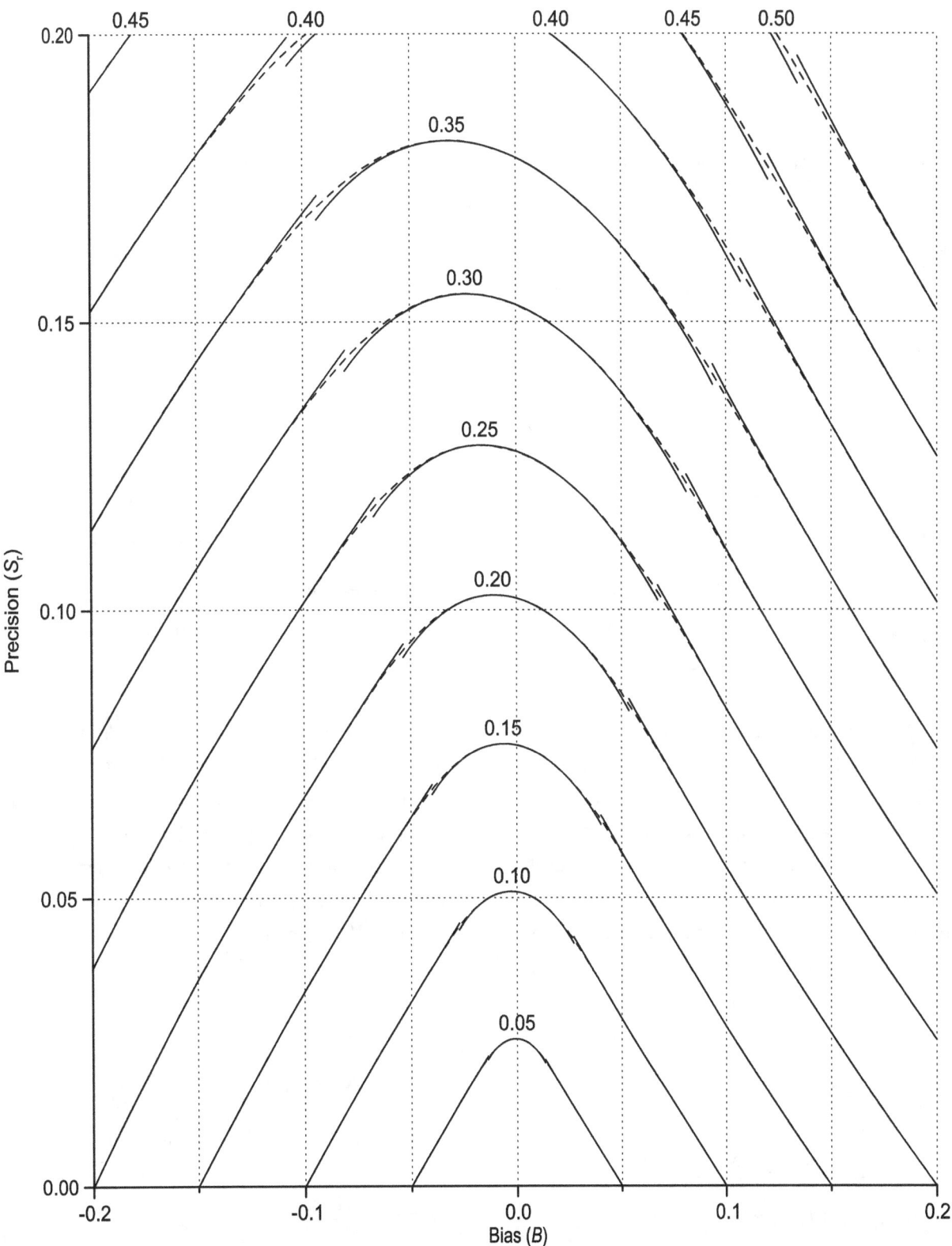

Figure A–2. Accuracy A as function of bias B and precision S_r by equation (A2) (solid lines) and equation (A1) (dashed lines). Each curve shows all values of B and S_r yielding A as indicated on the curve. These curves are similar to those in Figure 1 (page 21) of a previous evaluation protocol [NIOSH 1995], but S_r in this appendix is denoted by S_{rT} in that Figure 1.

Components for Evaluation of Direct Reading Monitors

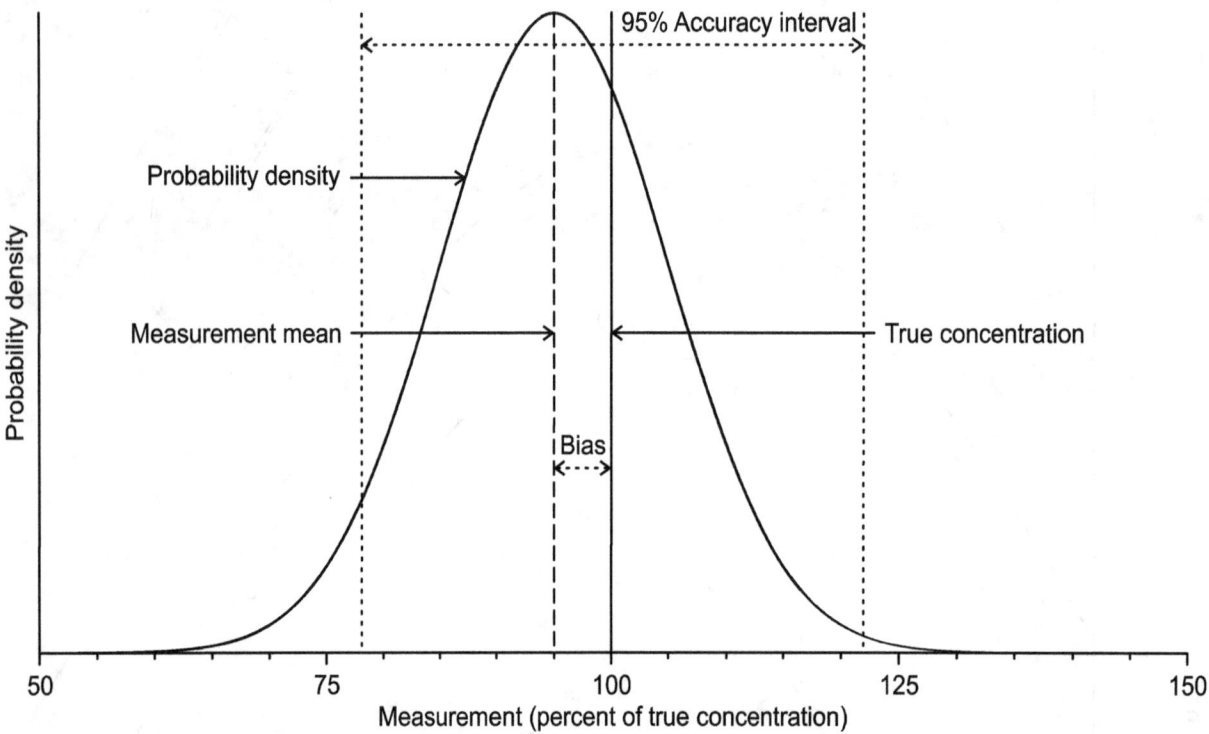

Figure A–3. Ninety-five percent accuracy interval by equation (A2) for a future measurement at given true concentration with $B = -0.05$ and $S_{rT} = 0.1$.

To estimate accuracy, bias and precision need to be estimated first. Suppose that $\hat{B}$ is an estimate of bias based on N data points; $\hat{S}_{rT}$ is an estimate of precision with M degrees of freedom. For the evaluation design using n samples from each of k concentrations, $N = n \times k$, $M = k(n-1)$. A point estimate of accuracy can be obtained by simply replacing the bias and precision parameters in the previous formula (A2) with their estimates as follows:

$$\hat{A} = A\left(\hat{B}, \hat{S}_{rT}\right)$$
$$= \begin{cases} 1.96 \times \sqrt{\hat{B}^2 + \hat{S}_{rT}^2} & \text{if } |\hat{B}| < \dfrac{\hat{S}_{rT}}{1.645}, \\ |\hat{B}| + 1.645 \times \hat{S}_{rT} & \text{otherwise} \end{cases} \quad \text{(A3)}$$

[AppA03].

When bias is homogeneous (see Appendix B), equation (A3) can provide a point estimate of the overall monitor accuracy over the range of concen-

trations using pooled values for bias $\hat{B}$ and precision ($\hat{S}_{rT}$ or $\hat{S}_r$). This can be calculated regardless of whether precision is homogeneous, since a *worst-case* precision (see Appendix B) can be estimated for all concentration levels, though other alternatives will be presented later.

A $(p \times 100)\%$ confidence limit estimate of accuracy for normally distributed measurements is given by

$$\hat{A}_p = \begin{cases} 1.96 \times \lambda \times \sqrt{\hat{B}^2 + \hat{S}_{rT}^2} & \text{if } |\hat{B}| < \dfrac{\hat{S}_{rT}}{1.645}, \\ |\hat{B}| + 1.645(\tau)\hat{S}_{rT} & \text{otherwise} \end{cases} \quad \text{(A4)}$$

[AppA04], where

$$\lambda = \sqrt{M / \chi^2_{1-p,M}},$$
$$\tau = t_{p,M}(\Delta)/\Delta,$$
$$\Delta = 1.645 \times \sqrt{N}$$

[AppA05].

Components for Evaluation of Direct Reading Monitors

$\chi^2_{1-p,M}$ is the $(1-p)\times100$ percentile of a chi-square distribution with M degrees of freedom (based on the Smith-Satterthwaite-Welch approximation [Welch 1956]), and $t_{p,M}(\Delta)$ is the $p\times100$ percentile of a noncentral t-distribution with M degrees of freedom and a noncentrality parameter Δ. (A more exact formula for λ is

$$\lambda = \sqrt{v/\chi^2_{1-p,v}},$$

$$v = \left[\hat{B}^2 + \hat{S}^2_{rT}\right]^2 \Big/ \left\{\left[(2/N)\hat{B}^2 + \hat{S}^2_{rT}/M\right]\hat{S}^2_{rT}\right\}$$

[AppA06]. However, unless the value of n is small, say, equal to 2, there is little difference if $v = M$.)

The above equation (A4) can be used to construct confidence intervals for accuracy. For example, a 90% confidence interval estimate of accuracy is given by $\left(\hat{A}_{0.05}, \hat{A}_{0.95}\right)$. Since the goal is often to provide the upper limit with 95% confidence, $\hat{A}_{0.95}$ is usually of interest.

Equation (A4) can be used for several different forms of the estimate. The simplest form is that the target concentrations at which the evaluation experiment is done are known exactly. In addition, there are situations when the target is unknown and must be estimated, which also includes the situation where the target concentration is only known up to its relative error. Equation (A4) may be used for each of these situations by substituting the estimate of the target concentration or the target concentration itself for the true concentration. (Use computer code (a) in Section 3 of this appendix.)

The recommendation is to correct for the bias if the estimated bias magnitude exceeds 10%. The bias correction is implemented by dividing future observations by $(1+\hat{B})$, where this quantity is determined as the ratio of the test method mean to either the target concentration or the mean of an independent method. Use of this correction requires modification of the accuracy equation (A4). First consider the situation in which an independent method is used to estimate the experimental

concentration; its measurements are assumed to be normally distributed and the bias of the independent method is known from previous experiments. It is assumed that the independent method has been corrected for bias, if it is not negligible. Its precision may be either estimated or known. An average value is taken for each method, so that $(1+\hat{B})$ is the ratio of the average test unit determinations to the average of the independent method determinations. After bias correction, equation (A3) is modified:

$$\hat{A} = 1.96u_c$$
$$= 1.96\sqrt{(1+1/s)\hat{S}^2_{rT}\Big/\left(1+\hat{B}\right)^2 + \hat{S}^2_{rT_{Ref,Avg}}} \quad (A5)$$
$$= 1.96\sqrt{(1+1/s)\hat{S}^2_r + \hat{S}^2_{rT_{Ref,Avg}}}$$

[AppA07], where s is the number of statistically independent values averaged to estimate the bias. $\hat{S}^2_{rT_{Ref,Avg}}$ is an estimate of the independent method relative variance for the average of m independent method measurements, $\hat{S}^2_{rT_{Ref}}\big/m$. (If the relative variance is known, that value can be used.)

Note that equation (A5) is consistent with combined uncertainty [ISO GUM 1995] as the root-sum-of-squares of bias correction uncertainty and random uncertainty components. The resulting expression is a form of prediction interval.

In many instances, there will have been previous evaluations in which the bias and precision of the independent method will have been determined. If the bias is nonnegligible, the bias will be assumed to be corrected for, so that the independent method determinations are approximately unbiased. In this case, a $p\times100\%$ confidence limit for accuracy of the bias-corrected future measurement is given by extending results in Bartley [2011]:

$$A_p = \hat{A}\sqrt{\dfrac{\dfrac{v}{\chi^2_{1-p,v}}}{\sqrt{\left(1+\dfrac{2}{s^2} + \dfrac{2\hat{S}^4_{rT,Ref,Avg}}{S^4_r} + \dfrac{5}{s}\dfrac{\hat{S}^2_{rT,Ref,Avg}}{S^2_r}\right)}}} \quad (A6a)$$

[AppA08], where

$$v = \cfrac{u_c^4}{\left\{ \hat{S}_{rT,Ref,Avg}^4 + \cfrac{2}{s}\hat{S}_{rT,Ref,Avg}^2 \hat{S}_r^2 \\ + \left[\cfrac{1}{s-1}\left(1+\cfrac{1}{s}\right)^2 + \cfrac{1}{s^2} \right] \hat{S}_r^4 \right\}}$$ (A6b)

[AppA09]. (Use computer code (b) in Section 3 of this appendix.)

Equations (A5) and (A6) can also be applied to other situations, with the following modifications.

(1) If the comparison data are collected in sets (n test unit(s), m independent method unit(s)), then $(1+\hat{B})$ is the average of the k ratios of test unit mean to independent method mean. $\hat{S}_r^2$ of (A6) is the average of $\hat{S}_{r,k}^2$ computed at each level k. $\hat{S}_{rT_{Ref,Avg}}^2$ is the independent method $\hat{S}_{rT_{Ref}}^2/km$. In place of s in (A6), use $N=kn$; for ($s-1$), use $k(n-1)$. This situation refers to situations 2 and 4 in the introduction to Appendix B.

(2) If there is no independent method and the target concentration is treated as known, then in equations (A5) and (A6), $\hat{S}_{rT_{Ref,Avg}}^2$ is replaced by 0. This refers to situation 1 in the introduction to Appendix B.

(3) If there is no independent method and the target has a relative standard deviation associated with it, $\hat{S}_{rT_{Ref,Avg}}^2$ is that value squared. This refers to situation 3 in the introduction to Appendix B.

There are instances where the independent method may not have been evaluated for bias, but for which a maximum bias Δ_{max} can be specified. For such cases, the following equations (A7) may be used [Bartley 2011], in which most symbols have the same meaning as in equations (A5) and (A6). Below, there is assumed to be just one independent method measurement. An estimate and a 95% confidence limit for accuracy are given in (A7):

$$\hat{A} = 1.96u_c$$
$$= 1.96\sqrt{\frac{1}{3}\Delta_{max}^2 + \left(1+\frac{1}{s}\right)\hat{S}_r^2}$$ (A7a)

[AppA10], and

$$A_p = \hat{A}\sqrt{\cfrac{\cfrac{v}{\chi_{1-p,v}^2}}{1+\cfrac{2}{s^2}+\cfrac{4}{45}\cfrac{\Delta_{max}^4}{S_r^4}+\cfrac{5}{s}\cfrac{\frac{1}{3}\Delta_{max}^2}{S_r^2}}}$$ (A7b)

[AppA11], where

$$v = \cfrac{u_c^4}{\left\{ \cfrac{2}{45}\Delta_{max}^4 + \cfrac{2}{3s}\Delta_{max}^2 \hat{S}_r^2 \\ + \left[\cfrac{1}{s-1}\left(1+\cfrac{1}{s}\right)^2 + \cfrac{1}{s^2} \right] \hat{S}_r^4 \right\}}$$ (A7c)

[AppA12]. (Use computer code (c) in Section 3 of this appendix.)

If bias is not homogeneous, then the above equations can be used to calculate point estimates and confidence limits for accuracy at individual concentration levels. Based on these individual concentration levels, a statement about accuracy can be made that the accuracy of the monitor is no worse than the highest value for accuracy calculated for the individual concentration levels, though an alternative will be presented in Appendix B.

As was discussed in Part I, Definitions, Bias, the calibrants used with the monitors have uncertainties, provided by the suppliers. The effect of the uncertainty on the accuracy estimate depends on its size relative to other sources of error, such as the bias in the monitor units under study and the bias in the independent method used in the evaluation.

The above formulas for the confidence interval are specifically for the evaluation described: n

Components for Evaluation of Direct Reading Monitors

determinations of a single unit at each of k levels. Modifications of this design will require modifications of equation (A4). For example, suppose that several monitor units are evaluated, and that, in addition to evaluating each unit individually, accuracy must be assessed for all monitor units combined, treating the units as a random sample from a larger population. For this application, interunit variability must be included in the total relative standard deviation (see Appendix C). In some evaluations there may be interest in day-to-day variability of monitors. This is also an extra component that must be included in the total relative standard deviation. When extra relative standard deviation components are included, beyond a single component for the test monitor and a single component for the independent method (or for the chamber where the evaluation takes place), the recommendation is to use equations (A3) and (A4) with $\hat{S}_{rT}^2$ equal to the total of the relative standard deviations squared.

3. R Code [R Project 2011] for Equations (A4), (A6), and (A7)

(a) Equation (A4)—no correction for bias in test method

```
# p is the confidence level
p <- 0.95
k <- 4
n <- 6
N <- k * n
M <- k * (n - 1)
# bhat denotes estimated bias
bhat <- 0.08
# srt denotes the estimated standard deviation relative to the true
#    concentration
srt <- 0.1
lam <- sqrt(M / qchisq(1 - p, M))
del <- 1.645 * sqrt(N)
tau <- qt(p, M, del) / del
A_hat <- ifelse(abs(bhat) < srt / 1.645, 1.96 * sqrt(bhat^2 + srt^2),
                abs(bhat) + 1.645 * srt)
Ap <- ifelse(abs(bhat) < srt / 1.645, 1.96 * lam * sqrt(bhat^2 + srt^2),
             abs(bhat) + 1.645 * tau * srt)
```

(b) Equations (A6)—test method assumed to be corrected for bias

```
p <- 0.95
k <- 2
n <- 6
s <- k * n
# sr_hat denotes estimated test method relative standard deviation
sr_hat <- 0.1
# srt_ind denotes the estimated independent method standard deviation
#    relative to the true concentration
srt_ind <- 0.1
# m = number of independent method measurements
m <- 6
df <- k * (n - 1)
div <- 1 + 2 / s^2 + 2 * (srt_ind^2 / m)^2 / sr_hat^4 + 5 / s *
  (srt_ind^2 / m) / sr_hat^2
```

```
mult <- (srt_ind^2 / m)^2 + 2 * (srt_ind^2 / m) * sr_hat^2 / s + sr_hat^4 *
  (1 / (df) * (1 + 1 / s)^2 + 1 / s^2)
A_hat <- 1.96 * sqrt((1 + 1 / s) * sr_hat^2 + srt_ind^2 / m)
# nu below is from equation (A6b)
nu <- (A_hat / 1.96)^4 / mult
# Ap is from equation (A6a)
Ap <- A_hat * sqrt((nu / qchisq(1 - p, nu)) / div)
```

(c) Equations (A7)—maximum bias available for independent method

```
p <- 0.95
k <- 3
n <- 6
s <- k * n
del_max <- 0.05
# sr_hat denotes estimated test method relative standard deviation
sr_hat <- 0.1
df <- k * (n - 1)
# del_max is maximum value for independent method
srt_ind <- sqrt(0.3333 * del_max^2)
div <- 1 + 2 / s^2 + (4 / 45 * del_max^4) / sr_hat^4 + 5 / s *
  (srt_ind^2) / sr_hat^2
mult <- 2 / 45 * del_max^4 + 2 * srt_ind^2 * sr_hat^2 / s + sr_hat^4 *
  (1 / (df) * (1 + 1 / s)^2 + 1 / s^2)
# A_hat is estimated accuracy, eq (A7a)
A_hat <- 1.96 * sqrt((1 + 1 / s) * sr_hat^2 + srt_ind^2)
# nu is the estimated degrees of freedom for chi square, eq (A7c)
nu <- (A_hat / 1.96)^4 / mult
# Ap below is the upper 100p% confidence limit for accuracy, eq (A7b)
Ap <- A_hat * sqrt((nu / qchisq(1 - p, nu)) / div)
```

4. References

Bartley DL [2001]. Definition and assessment of sampling and analytical accuracy. Ann Occup Hyg 45(5):357–364.

Bartley DL [2011]. Confidence level on accuracy of a calibrated method. Task order no. 2–2011 for CDC/NIOSH/DART project, Statistical Research, Development and Planning. Unpublished.

ISO GUM [1995]. Guide to the expression of uncertainty in measurement, ISO Guide 98. Geneva: International Organization for Standardization.

Krishnamoorthy K, Mathew T [2009]. Inference on the symmetric-range accuracy. Ann Occup Hyg 53(2):167–171.

NIOSH [1995]. Guidelines for air sampling and analytical method development and evaluation. By Kennedy ER, Fischbach TJ, Song R, Eller PM, Shulman SA. Cincinnati, OH: U.S. Department of Health and Human Services, Centers for Disease Control and Prevention, National Institute for Occupational Safety and Health, DHHS (NIOSH) Publication No. 95–117. [http://www.cdc.gov/niosh/docs/95-117/pdfs/95-117.pdf]

R Project [2011]. The R project for statistical computing [http://www.r-project.org/]. Date accessed: November 2011.

Welch BL [1956]. On linear combinations of several variances. J Am Stat Assoc 51:132–148.

Appendix B. Statistical Evaluation of Bias and Precision for Individual Monitor Units

1. Introduction

The performance of a direct-reading monitor is determined by its response behavior. Assume that at any given sample concentration, the monitor responses are normally distributed. Under this assumption, the response distribution can be characterized by its mean and standard deviation. Let C_T denote the true sample concentration, and let μ and σ be the response mean and standard deviation. Then the bias, denoted by B, is defined as the relative difference from the true concentration,

$$B = \mu/C_T - 1. \tag{B1}$$

The precision is measured by the relative standard deviation,

$$S_r = \sigma/\mu, \tag{B2}$$

or the standard deviation relative to the true concentration,

$$S_{rT} = \sigma/C_T = (1+B) \times S_r. \tag{B3}$$

Both bias and precision may vary as sample concentration changes. Therefore, the homogeneity of bias and the homogeneity of precision need to be evaluated by samples from at least four concentrations, covering the concentration range considered. If differences are insignificant, then pooled estimates of bias and precision should be derived; otherwise, separate estimates must be obtained at each concentration level, or for levels where bias and precision are poolable. Suppose that n samples from each of the k concentrations are used in the statistical evaluation. Response results are denoted

by x_{ij} for $i = 1, \ldots, k$ and $j = 1, \ldots, n$. The true concentration and the mean and standard deviation of response results at each concentration are denoted by C_{Ti}, μ_i, and σ_i, respectively, for $i = 1, \ldots, k$. Bias and precision at each level are accordingly defined and denoted by B_i (B1) and S_{ri} (B2) or S_{rTi} (B3).

For evaluation of either homogeneity of bias or of precision, it is necessary to inspect data for outliers, which can both increase estimated variability and affect bias estimates. Although obvious outliers can be apparent from plots of the data by concentration level, use of statistical tests can be helpful for identification of potential outliers. The easiest to use, such as Grubb's test, can be used for the n determinations at each level [Barnett and Lewis 1978].

True concentrations of test samples are needed to determine bias. However, true concentrations may not be known without error. If certified standard materials are used, then the reference values provided are estimates of true concentrations, and the standard uncertainty values are the standard deviations of the estimates of the reference values. If field samples are used or if a generation chamber is used that does not adequately control the concentration, then an independent method is needed to estimate true concentrations. In this case, m samples from each of k concentrations are analyzed by the independent method to establish the reference values and associated uncertainty values. In the discussion in this appendix, it is assumed either 1) the independent method samples are statistically independent or 2) if the independent method uses multiple determinations by multiple monitor units, then the differences between units are assumed small compared to the measurement error of each unit's determinations. If the independent method is used as described in situation (2) and monitor unit differences are not small, then the methods of Appendix C must be used to evaluate the direct-reading monitor unit bias, precision, and accuracy.

It is assumed that the test and independent method measurements are either normally distributed or, if specified, lognormally distributed. The only exception is when there are no data for the independent method (see Appendix A, equation (A7)).

The approach presented here assumes that there are multiple determinations by the monitor unit under study at each concentration, except for situation 4 below. In order to estimate bias and precision, several different kinds of sampling situations must be noted. For each of these situations, the study monitor unit must be reinitialized, so that each determination may be regarded as statistically independent of every other determination. Situations 1, 2, and 3 are based on laboratory experiments. Situation 4 would most likely be a field situation.

(1) The true concentration is known. For this case, a design can be used in which the order of the various concentrations at which the unit is tested is randomized. If this is too difficult to do, then successive determinations at the same concentration can be used.

(2) The true concentration is not known, but it can be maintained over a substantial period of time, so that replicates of the unit under evaluation can be obtained at that concentration. Successive determinations at each concentration level is the appropriate procedure to use here, rather than the random order suggested for 1) above. An independent method must be used to estimate the concentration. It will usually be assumed that the independent method is unbiased or has negligible bias. (See additional discus-

sion at the end of this section.) However, there are cases where this may not be true. For instance, biased independent methods are sometimes used as reference methods, in which case determination of bias relative to the reference method has practical importance.

(3) The true concentration is known by stoichiometric calculations, which means that the target concentration can be calculated, but the actual concentration is assumed to differ from the target, usually by a normally distributed random variable, with mean 0.

(4) The true concentration is not known and it is difficult to control the concentration to which the monitor unit is exposed, or the experimenters have made only one evaluation of the test method at each concentration. An independent method must be used to estimate the concentration.

As was discussed in Appendix A, it is possible that calibrant error can affect the estimate of bias and the precision of the bias estimate. The effect will vary by the type of the data. For sampling situation 1, for which the true concentration is known, there will be no calibrant error in the generation of the test concentration, but the calibrant error will affect the unit response. For situations 2 and type 4 data, if the test unit and the independent method are calibrated from the same calibrant, then there will be no effect of calibrant error, since in the ratio of test unit to independent unit measurements, calibrant error will be removed. If different calibrants are used, this does not happen. In data from situation 3, where calibrant error affects both the unit itself and the generated test concentrations, the generated test concentration calibrant is analogous to the independent method calibrant, and analogous remarks apply as for the data from situations 2 and 4.

To understand the effect of a different calibrant source, information about calibrant uncertainty must be obtained. Most suppliers provide a concentration with a plus/minus value in the data sheet accompanying the calibrant, usually given in percent. The true concentration is thought to lie between L = (specified concentration) × u_L and U = (specified concentration) × u_U, where u_L = 1 minus (uncertainty value) and u_U = 1 plus (uncertainty value). A statistician should be consulted if the uncertainty in calibrant bias seems large enough to warrant concern.

In the following sections, the intent is to evaluate the bias and precision of the test method, not of the independent method. Precision of the independent method can be estimated in the same experiment in which the test method is evaluated. However, bias requires past evaluation. If there is past evaluation, then the independent method can be corrected for the bias, if there is need to correct. If there are no past data, there can be no correction, but a technique is presented in Appendix A for placing bounds on the bias, and thereby computing accuracy.

2. Evaluation of Bias

The bias evaluation includes (1) the homogeneity test of bias and (2) the estimation of bias. If there is no significant evidence to reject the hypothesis of homogenous bias, a pooled estimate of bias should be obtained. A bias correction should be considered if the magnitude of the (pooled) bias estimate is greater than 10%. Since the need to correct for the bias is based on the sample value, no test is presented that the pooled bias exceeds 10%.

2.1. Homogeneity test of bias, true concentration known

In the case that true concentrations of test samples are known without error, the homogeneity of bias

should be tested using the F-test in the analysis of variance (ANOVA). This test uses the ratio of sample determinations to the true concentration, the simplest situation. The ANOVA procedures assume homogeneous variance. When the true concentration in a sample group is known and is used as the divisor, then the standard deviation of the ratios is the S_{rT} of the test method results. It makes sense to test first for homogeneity of the standard deviation of the ratios described in Section 3.1 of this appendix. The ANOVA procedure for ordinary least squares is as follows:

2.1.1. Compute the sum of squares within each sample group:

$$SS_w(y) = \sum_{i=1}^{k}\sum_{j=1}^{n}\left(y_{ij} - \bar{y}_{i.}\right)^2,$$

$$\bar{y}_{i.} = \frac{1}{n}\sum_{j=1}^{n}y_{ij}, y_{ij} = \frac{x_{ij}}{C_{Ti}} - 1 \tag{B4}$$

[AppB01].

2.1.2. Compute the sum of squares between sample groups:

$$SS_b(y) = \sum_{i=1}^{k}n\left(\bar{y}_{i.} - \bar{y}_{..}\right)^2, \bar{y}_{..} = \frac{1}{k}\sum_{i=1}^{k}\bar{y}_{i.} \tag{B5}$$

[AppB02].

2.1.3. Calculate the F-ratio:

$$F = \frac{SS_b(y)/(k-1)}{SS_w(y)/(k(n-1))} \tag{B6}$$

[AppB03]. Under the null hypothesis, this ratio should be an observed value from an F-distributed variable with degrees of freedoms $(k-1)$ and $k(n-1)$.

2.1.4. Select a significance level α (e.g., $\alpha = 0.05$).

2.1.5. Find the probability that an F-distributed variable with degrees of freedoms $(k-1)$ and $k(n-1)$ has a value greater than the observed ratio F (B6). This probability is called the p-value of the test and can be obtained from an F-table or from most computer programs. Reject the null hypothesis that the bias is constant if and only if the p-value is less than the significance level. Alternatively, one can obtain a critical F-value for the degrees of freedom, choose α from an F-table, and reject the null hypothesis if and only if the F-value (B6) exceeds the critical value of F.

A common reason that a monitor fails the homogeneity test is the presence of a trend in bias over the concentration levels tested. Statistical significance of an apparent trend can be determined by regression analysis. The presence or absence of a statistically significant trend should be noted.

This test can be performed by an analysis of variance procedure using C_{Ti} (known concentration levels) as the class variable. Most statistical packages that have analysis of variance programs allow for the use of weighted least squares, which is appropriate if S_{rT} is not constant. The weights to be used are the reciprocals of the estimated S_{rT}^2 values, though it may be possible to group estimated S_{rT}^2 values into poolable groups, in which case the reciprocals of the pooled S_{rT}^2 values are used as weights for the selected groups (see Section 3 of this appendix).

2.2. Estimation of bias

2.2.1. Bias homogeneous

If the bias is homogenous, a pooled estimate of bias should be derived from equation (B5) as

$$\hat{B} = \overline{y}_{..}. \tag{B7}$$

The ratio of sample determinations is relative to either the independent method determination or the true concentration in the sample group.

2.2.2. Bias not homogeneous

If the homogeneity test (B6) is failed, then the simplest procedure is to evaluate bias at each concentration [see equation (B4)] by

$$\hat{B}_i = \overline{y}_{i.}.$$

A more complicated, but perhaps more useful, procedure is to identify levels for which the bias is poolable. This requires use of multiple comparison tests, such as the Tukey multiple comparison procedure [Miller 1981a]. It may be that concentration levels can be pooled several different ways, since the lowest bias(es) and the highest bias(es) may be statistically distinguishable from each other, but bias estimates in between these may or may not be distinguishable from either the lowest or highest. If an estimated bias of magnitude greater than 10% can be pooled with estimated bias(es) of less than 10% magnitude, then it is possible that the pooled value will be less than 10% and will satisfy the bias criterion. If the bias estimates are not poolable, but all are less than 10% magnitude, then a conservative estimate of bias is the bias with the largest absolute value. If there is a trend in bias over concentration levels, then the homogeneous groups can be obtained by ordering the concentrations and dividing the ordered concentrations into several groups.

2.2.3 Bias relative to what?

Of the four designs specified in Section 1 above, designs (1) and (3) offer either the true concentration value for each group or an unbiased estimate of the true concentration. Designs (2) and (4) rely on an independent method to provide estimates of bias. For some independent methods the bias and precision may be unknown. If the independent method is a reference method, then the there may be interest in bias relative to that reference method, even if the reference method is biased. However, there may be many cases where the interest is in determining bias of the test method relative to the true concentration. There are several cases to consider:

(1) Bias and precision of the independent method are known or estimated. Bias of the independent method is estimated by the ratio of independent method to true mean concentrations via a prior evaluation. Future independent method bias is corrected by dividing a future measurement by this ratio. With or without independent method bias correction, the variance of its measurements is included in equation (A5), Appendix A, because correction of the test method for its bias relative to the independent method includes independent method variability.

(2) Bias and precision of the independent method are neither known nor estimated. The accuracy computations for this case are considered in equation (A7) of Appendix A.

2.3. Homogeneity test of bias: true concentrations are unknown, but can be maintained, or a target concentration is used as the true concentration

When the concentration is unknown, the known concentration used in the denominator of y_{ij} must be replaced by the estimates of the true values, usually those produced by an independent sampling method. Alternatively, there are situations where a target concentration may be used, even though the true concentration will differ from the target, for example, when the target is calculated by

stoichiometric computations. We assume that approximately the same concentration is maintained for each unit in trials at that concentration. If so, then it makes little difference for the first situation whether test and independent method units are evaluated simultaneously or separately. For either of these two situations, the homogeneity test of bias will have higher than expected type I error, which means that it will too often reject the hypothesis of constant bias when the hypothesis is true. The reason for this problem is that measurements y_{ij} at the same concentration are positively correlated, and the variance is underestimated. To correct the problem, the following F'-test statistic (B8) [Song et al. 2001] should be used in the test procedure:

$$F' = \frac{SS_b(y')/(k-1)}{SS_w(y')/(k(n-1)) + (n/m)\hat{\sigma}_R^2} \quad (B8)$$

[AppB04], where $SS_b(y')$ and $SS_w(y')$ [(B5) and (B4)] are calculated using

$$y'_{ij} = \left\{ x_{ij} / \left[\hat{C}_{Ti}(1+\hat{B}) \right] \right\} - 1$$

in place of y_{ij}, $\hat{C}_{Ti}$ is either the estimate of C_{Ti} or the target value, $\hat{B}$ is the pooled estimate of bias (B7), $\hat{\sigma}_R^2$ (B9a) is the estimate of the pooled squared relative standard deviation of $\hat{C}_{Ti}$, and the F'-value is to be compared to the 95% value of an F-statistic with $(k-1)$ and $k(n-1)$ degrees of freedom. If the value of F' exceeds the 95% value, then the hypothesis of bias homogeneity is rejected.

Note well that the y'_{ij} values are calculated by dividing the x_{ij} by the product of (1 + average test method bias over all levels) and the estimated true concentration for level i, or the target concentration. (The F'-test statistic (B8) differs from that in the reference [Song et al. 2001] as follows. The ratio of measurement to estimated concentration is used here rather than the natural log of that ratio used in the reference. Because of the change to the original scale, the ratio must be divided by the term involving the pooled bias. Type I errors are similar for either scale.) If an independent method is used to estimate the true concentrations, and m samples from each of k concentrations are analyzed by the independent method, average the m estimates $\{c_{ij}\}$ at each concentration i:

$$\hat{C}_{\mathrm{T}i} = \frac{1}{m}\sum_{j=1}^{m} c_{ij}, \text{ and also}$$

$$\hat{\sigma}_{\mathrm{R}}^2 = \frac{1}{k(m-1)}\sum_{i=1}^{k}\sum_{j=1}^{m}\left(c_{ij} - \hat{C}_{\mathrm{T}i}\right)^2 / \hat{C}_{\mathrm{T}i}^2 \tag{B9a}$$

[AppB05]. If the target concentration is used, then the variance associated with the target concentration should be used in place of that given in (B9a). Also, $m = 1$ for this case. If the bias is found to be homogeneous, then follow the procedure under Section 2.2.1 in this appendix. If the bias is determined not to be homogeneous, follow the procedures of Section 2.2.2 in this appendix. (Notice that $\hat{S}_{ri}^2(\hat{C}_{\mathrm{T}i}) = \sigma_{\mathrm{R}}^2 / m$.)

When the test (B8) is used with independent method estimation of the concentration, its use requires that the S_{rT} values of the test method and independent method be poolable over the concentration levels to which the test is applied. (Alternatively, the test also can be used for levels over which the ratio of test method to independent method is constant, but it would be difficult to determine whether this is true.) If this is not so, then the simplest approach is perhaps to use the methods of Section 3.2 in this appendix to determine poolable levels, and apply the F′-test to the determinations from those levels. If there were several groups of poolable levels, this would require an F′-test for each group. The test (B8) does not require that the independent method be unbiased, although it does require homogeneity of bias over the concentration levels.

2.4. Homogeneity of bias: true concentrations are not known, and it is difficult to control the concentration in the chamber; for each trial the test unit and at least one independent method unit are evaluated

In this section, the notation is presented in terms of the unnormalized values $x_{ij(\text{test})}$ and $x_{ij(\text{ind})}$ for the test method and independent method, respectively. Let $z_{ij(\text{test})} = \ln x_{ij(\text{test})}$ and $z_{ij(\text{ind})} = \ln x_{ij(\text{ind})}$. The bias estimate is $\hat{B} = \exp[\overline{z}_{\cdot\cdot(\text{test})} - \overline{z}_{\cdot\cdot(\text{ind})}] - 1$, where the dot subscripts indicate averaging, analogous to that defined in (B4) and (B5). A difference from the previous evaluation situations is that the replications indexed by j within concentration i must all occur simultaneously, since the same concentration cannot be repeatedly generated in a reliable manner. For the test method, $j = 1$. For the independent method, j must be greater than 1 when the variance of the independent method is unknown. However, if $j = 1$ for the independent method, homogeneity of bias can be tested, but the variance of the independent method must be known if the variance of the test method is to be appropriately estimated by means of analysis of variance, as is done here. Let m denote the number of independent method replications at each trial, k the number of trials during the day. Make the following definitions:

$$\hat{\mu} = \frac{\sum_{i=1}^{k}\left(z_{i1(\text{test})} + m\overline{z}_{i\cdot(\text{ind})}\right)}{(1+m)k}, \quad \hat{\mu}_{\text{test}} = \frac{\sum_{i=1}^{k} z_{i1(\text{test})}}{k}$$

[AppB06],

$$\hat{\mu}_{\text{ind}} = \frac{\sum_{i=1}^{k} m\overline{z}_{i\cdot(\text{ind})}}{mk}, \quad \hat{\mu}_i = \frac{z_{i1(\text{test})} + m\overline{z}_{i\cdot(\text{ind})}}{1+m}$$

[AppB07],

$$\hat{\alpha}_{\text{test}} = \hat{\mu}_{\text{test}} - \hat{\mu}, \quad \hat{\alpha}_{\text{ind}} = \hat{\mu}_{\text{ind}} - \hat{\mu}, \quad \hat{\beta}_i = \hat{\mu}_i - \hat{\mu}$$

[AppB08],

Components for Evaluation of Direct Reading Monitors

$$\hat{\sigma}_{bias} = \sqrt{\frac{\left(\sqrt{\begin{array}{c} \sum_{i=1}^{k} z_{i1(test)}^2 + \sum_{i=1}^{k} m\bar{z}_{i\cdot(ind)}^2 \\ -\left[\begin{array}{c} k(1+m)\hat{\mu}^2 + k\left(\hat{\alpha}_{test}^2 + m\hat{\alpha}_{ind}^2\right) \\ +\sum_{i=1}^{k}(1+m)\hat{\beta}_i^2 \end{array}\right] \end{array}}\right)}{(k-1)}} \quad (B9b)$$

[AppB09], and

$$\hat{\sigma}_{res} = \sqrt{\frac{\sum_{i=1}^{k}\left[z_{ij(ind)} - \bar{z}_{i\cdot(ind)}\right]^2}{k(m-1)}} \quad (B9c)$$

[AppB10], where $\hat{\sigma}_{res}$ is the estimated precision of the residuals.

Under the hypothesis that the bias does not vary over the k trials and assuming that the variances are constant over the k trials and the variance of the independent method is not greatly different from that of the test method, the quantity $G = \hat{\sigma}_{bias}^2/\hat{\sigma}_{res}^2$ has a central F-distribution with $(k-1)$ and $k(m-1)$ degrees of freedom. If G is less than the 95th percentile of the F-distribution, then the assumption of homogeneity of bias is not contradicted by the data. Otherwise, accuracy should be estimated for each concentration. (Computer code to calculate and evaluate statistical significance of G is given in Section 7 of this appendix.) We assume that the variance of the test method does not differ much from that of the independent method. If the test method variance is much larger, say, more than two or three times larger, then the test will give significant results more often than it should, since the true variance will be larger than the expected value of the estimate. If the test method variance is much smaller, then the test will give significant results less often than it should. A sensible graphical procedure that would supplement the above test is to plot the test method values versus the average of the independent method values, together with the least squares fitted straight line. Large devia-

tions of actual values from the straight line could indicate inhomogeneous bias. This would support the results of the above test, when that test yields a statistically significant result. If the F-test does not yield a statistically significant result, then the $\hat{\sigma}_{bias}$ and $\hat{\sigma}_{res}$ can be squared and multiplied by their degrees of freedom. The sum of these two quantities is divided by the sum of the degrees of freedom to obtain a pooled estimate of the variance of measurement, yielding

$$\hat{\sigma}_{test}^2 = \frac{(k-1)\hat{\sigma}_{bias}^2 + k(m-1)\hat{\sigma}_{res}^2}{km-1}$$

[AppB11].

If the F-test yields a statistically significant result, then groups of concentrations must be determined in which bias is approximately homogeneous. A reasonable approach is to omit either the lowest or highest concentration measurements and carry out the procedure described above.

If there is no replication of the independent method but the relative standard deviation is provided by prior knowledge, then the relative standard deviation S_r can be used in place of $\hat{\sigma}_{res}$ in the computation of the statistic G. In this case, $(k-1)$ × modified G is compared to the 95th percentile of the chi-square distribution with $(k-1)$ degrees of freedom.

3. Evaluation of Precision

3.1. Homogeneity tests for precision, true concentration known, or the concentration can be maintained for repeated trials, or the concentration is known by stoichiometric calculations

For accuracy calculations, precision can be evaluated in terms of S_r or S_{rT}. Given the bias B, S_{rT} and S_r can be derived from each other by using the relationship $S_{rT} = (1+B) \times S_r$. If bias is homogeneous, then testing the homogeneity of S_{rT} is equivalent to testing the homogeneity of S_r. If both S_r and S_{rT} are homogeneous, then, of course, bias must be homogeneous. Also of interest is the homogeneity of the standard deviations of the ratio of sample determinations to the true concentration in the sample group. The homogeneity test presented below for S_{rT} values may be interpreted as a homogeneity test for the standard deviation of the ratios when the true concentrations are known.

The homogeneity of precision may be determined by Bartlett's test. To use Bartlett's test procedure [Bartlett 1937] for a homogeneity test of S_{rT}, use $y_{ij} = (x_{ij}/C_{Ti}) - 1$ or $y_{ij} = (x_{ij}/\hat{C}_{Ti}) - 1$. The first form above applies when the true value is known; the second form applies when the true concentration is estimated by an independent method or the target concentration is used. To use the second form, the assumption is made that the relative standard deviation of the independent method or of the true concentration relative to the target is either small compared to that of the test method or is homogeneous over concentration levels.

For the homogeneity test of S_r, use

$$y_{ij} = \left(x_{ij}/\overline{x}_i\right) - 1,$$

where

$$\overline{x}_i = \sum_{j=1}^{n} x_{ij}/n$$

[AppB12].

3.1.1. Calculate the test statistic

$$H = \frac{k(n-1)\ln\left(\hat{\sigma}_e^2\right) - \sum_{i=1}^{k}(n-1)\ln\left(\hat{\sigma}_i^2\right)}{1+(k+1)/\left[3k(n-1)\right]} \quad \text{(B10a)}$$

[AppB13], where

$$\hat{\sigma}_e^2 = \frac{1}{k}\sum_{i=1}^{k}\hat{\sigma}_i^2$$

[AppB14] and

$$\hat{\sigma}_i^2 = \frac{1}{n-1}\sum_{j=1}^{n}\left(y_{ij} - \overline{y}_i\right)^2$$

[AppB15]. $\hat{\sigma}_i^2$ is the estimated S_{rTi}^2 or S_{ri}^2 for level i. More general formulas are available if the number of samples in each concentration group is not the same [Hald 1952]:

$$H = \frac{N'\ln\left(\hat{\sigma}_e^2\right) - \sum_{i=1}^{k}(n_i - 1)\ln\left(\hat{\sigma}_i^2\right)}{1+\dfrac{1}{3(k-1)}\left[\left(\sum_{i=1}^{k}\dfrac{1}{n_i - 1}\right) - \dfrac{1}{N'}\right]} \quad \text{(B10b)}$$

[AppB16], where

$$\hat{\sigma}_e^2 = \sum_{i=1}^{k}(n_i - 1)\frac{\hat{\sigma}_i^2}{N'}$$

[AppB17], n_i is the number of measurements used to calculate $\hat{\sigma}_i^2$, and

$$N' = \sum_{i=1}^{k}(n_i - 1)$$

[AppB18]. Under the null hypothesis, the statistics (B10a) and (B10b) should be observed values from an approximately χ^2 distribution with $(k-1)$ degrees of freedom. However, the chi-square approximation is not good if many n_i values equal 2. Alternatively, when many n_i values equal 2, a test described by Miller can be used [Miller 1981b].

3.1.2. Select a significance level α (e.g., $\alpha = 0.05$).

3.1.3. Find the probability that a chi-square variable with $(k-1)$ degrees of freedom has a value greater than the observed value H (B10a or B10b).

Components for Evaluation of Direct Reading Monitors

This probability is the *p*-value of the test and can be obtained from the χ^2 table or from many computer programs. Reject the null hypothesis that the bias is constant if and only if the *p*-value is less than the significance level. Alternatively, one can obtain a critical χ^2 value for the degrees of freedom, choose α from a χ^2 table, and reject the null hypothesis if and only if the χ^2 value (B10a or B10b) exceeds the critical value of χ^2.

As was mentioned with regard to inhomogeneity of bias (Section 2.1.5 in this appendix), a common reason that a monitor fails the homogeneity test for precision is the presence of a trend in precision over the concentration levels tested. The presence or absence of a statistically significant trend should be noted.

3.2. Estimation of precision

3.2.1. Precision homogeneous

If S_{rT} or S_r is pooled, then the pooled estimate of precision is given by $\hat{S}_{rT}$ or $\hat{S}_r$:

$$\hat{S}_{rT} = \sqrt{\sum_{i=1}^{k} \frac{(n_i - 1) \times \left(\hat{S}_{rTi}\right)^2}{\sum_{l=1}^{k}(n_l - 1)}}$$

[AppB19],

$$\hat{S}_{r} = \sqrt{\sum_{i=1}^{k} \frac{(n_i - 1) \times \left(\hat{S}_{ri}\right)^2}{\sum_{l=1}^{k}(n_l - 1)}}$$

[AppB20]. The $\hat{S}_r$ can be converted to $\hat{S}_{rT}$ by multiplying $\hat{S}_r$ by $(1 + \hat{B})$ if the $\hat{B}_i$ values are poolable. When C_{Ti} is estimated (as discussed in Section 3.1), $\hat{S}_{rT}$ will include variability associated with the estimate. If that variability is not small relative to the variability of *x*, then a statistician should be consulted.

3.2.2. Precision not homogeneous

If Bartlett's test (B10a or B10b) indicates that the precision is not homogeneous, then Bartlett's test is to be applied to various groups of precision estimates from the concentration levels studied. Several different strategies can be used for reporting the monitor's S_{rT} or S_r, or for determining weights to be used in the weighted least squares model for bias (Section 2.2 in this appendix). The discussion below is given in terms of S_{rT} values.

(1) The concentration levels can be divided into groups, such that in each group, S_{rT} estimates are homogeneous and, therefore, poolable. If there is a trend in S_{rT} over concentration levels, then the homogeneous groups can be obtained by ordering the concentrations and dividing the ordered concentrations into several groups.

(2) It may often occur that inhomogeneity is found at the lowest concentrations. Thus, these lowest concentration(s) can be handled separately (either individually or by pooling them), and the S_{rT} values for the remaining concentrations can be pooled. For both weighted least squares and for reporting precision, the S_{rT} values can be those determined for the concentration groupings.

(3) If the S_{rT} estimates are not homogeneous, and if the highest estimate does not occur at the lowest concentrations, then pooling can still be attempted. (It seems important to conduct outlier tests before proceeding, since the presence of outliers can increase the S_{rT} estimates.) This situation is more complicated than (2) because the occurrence of the highest S_{rT} at higher concentrations is usually unexpected, and may be difficult to explain. Choices for pooling estimates into groups are:

(a) If there is a concentration level that is suspected of contributing to the S_{rT} inhomogeneity, then leave that level out of the

Bartlett's test calculation. If the remaining values are homogeneous, then calculate a pooled S_{rT} estimate based on these values. If the pooled S_{rT} value is greater than the S_{rT} that was excluded from the calculations, then use the pooled value in further calculations. (In some instances, more than one precision value may need to be excluded.) If the pooled value is less than the excluded value(s), then use the excluded S_{rT} value (largest value) as representative (as a worst case) of the precision of the measurements.

(b) The largest precision value can be combined with other precisions to see if these combinations are homogeneous. If so, then the largest pooled precision of these combinations can be used as the worst-case estimate of precision.

(c) Precision and the associated accuracy can be reported by the different groupings, without deciding on a worst-case estimate.

For example, the precisions S_{rTi} representing 6 replicates from 4 concentration levels are 0.02, 0.03, 0.07, and 0.01, listed by levels from lowest to highest concentrations. Bartlett's test (B10a) indicates that these precision estimates are not homogeneous, but values from the first, second and fourth concentration levels are homogeneous. The pooled precision from these values is 0.02. However, since this value is smaller than the excluded value of 0.07, then 0.07 might be used as the worst-case estimate of precision (choice (a) above). Alternatively, Bartlett's test can be applied to a group of precision estimates that includes the 0.07 value. In this situation, the second (0.03) and third (0.07) values are homogeneous according to Bartlett's test and can be pooled to give a worst-case estimate of 0.054 (choice (b) above). Choice (c) consists of reporting results for either the homogeneous groupings of choice (a) or (b). Choice (c) gives the

groupings to use for calculating the S_{rT} values for the weighted least squares (Section 2.1.5).

Another option is to use a test for homogeneity of sample variances based on the analysis of variance, as described in Scheffe [1959]. If this test is used, then multiple comparison methods (as was discussed for bias in Section 2.2.2 in this appendix) from the analysis of variance can be used to determine statistically significant differences. However, because of possible nonnormality of sample variances, simulations may be needed to determine if the nominal confidence limits are accurate.

3.3. True concentrations are not known, and it is difficult to control the concentration in the chamber; for each trial the test unit and at least one independent method unit are evaluated

As developed in Section 2.4 in this appendix, the approach relied on simultaneous evaluation of test method units and independent method units, with just one evaluation of the test method at each concentration. Because of the lack of replication of the test method, it is not possible to directly carry out Bartlett's test with the test method measurements. The graphical procedure suggested at the end of Section 2.4 could be helpful here, if there are enough trials and enough spread among the concentrations. Sets of consecutive plotted points that lie on one side of the fitted line may indicate bias for measurements at that concentration, and sets of points that form clusters closer or further from the line may indicate changes in variance. However, these kinds of distinctions are not always easy to make.

The pooled estimate $\hat{\sigma}^2_{test}$ of Section 2.4 can be used as an estimate of the relative variance of the test method unit. (Recall that the data are analyzed on the natural log scale). Alternatively, if a non-pooled estimate is required, the following approach can be used. If the independent method has known constant relative variance, denoted by $\sigma^2_{r(ind)}$, and if the bias is determined to be homogeneous, then the test unit relative variance can be estimated by

$$\hat{\sigma}^2_{r(test)} = \left[(m+1)/m\right]\left[\hat{\sigma}^2_{bias} - \sigma^2_{r(ind)}/(m+1)\right].$$

The square root of this quantity is the estimated relative standard deviation of the test method with $(k-1)$ degrees of freedom. If the independent method relative variance must be estimated, then the same relation applies with the pooled estimate $\hat{\sigma}^2_{res}$ of (B9c) used in place of $\sigma^2_{r(ind)}$. First the poolability of the independent method variances must be established, as in Section 3.1. If these variances are poolable, then the relative variance of the test method unit can be estimated by

$$\hat{\sigma}^2_{test,alt} = \left[(m+1)/m\right]\left[\hat{\sigma}^2_{bias} - \hat{\sigma}^2_{res}/(m+1)\right],$$

which has degrees of freedom, computed by Welch-Satterthwaite's [Welch 1956] approximation:

$$\nu_{test,alt} = \frac{\left\{\left[(m+1)/m\right]\hat{\sigma}^2_{bias} - \hat{\sigma}^2_{res}/m\right\}^2}{\dfrac{\left\{\left[(m+1)/m\right]\hat{\sigma}^2_{bias}\right\}^2}{(k-1)} + \dfrac{\left(\hat{\sigma}^2_{res}/m\right)^2}{k(m-1)}}$$

[AppB21].

If the independent method does not have poolable precision, then poolable levels must be determined and the calculations can be carried out for levels that are poolable. When precision and bias are poolable, the square root of the relative variance $\times (1+\hat{B})^2$ is the estimated S_{rT}.

4. Evaluation of Precision over Time, Including Day-to-Day Variation

In the following sections, change in monitor response over days is treated as random. For some monitors, it is possible that there can be trends in average monitor response over time. This is something to be aware of, but is not discussed here.

4.1. True concentration known, or the concentration can be maintained for repeated trials, or the concentration is known by stoichiometric calculations

The total S_{rT} of measurements should include components associated with variation between and within days. To estimate these components, a monitor needs to be tested under conditions that are typical in the workplace. Tests must be performed several times each day and repeated on several consecutive days (see Part III, Performance Characteristics). The experiment proposed here is made with the assumption that the S_{rT} is constant for the concentration levels included in the experiment.

Let x_{ij} be the normalized test result at time t_j of the ith day, $i = 1,\ldots,d$, $j = 1,\ldots,n$. Here, a normalized result is a ratio of the actually measured result to the true concentration value of the substance being measured, or to the estimate of that concentration or to the target concentration, but indexed in terms of d days and n replications per day. The proposed evaluation is to be done at a single concentration, any concentration in the range under consideration for which the S_{rT} has been determined to be homogeneous, as determined by Section 3.1. For statistical independence, we recommend that the evaluation of bias be done separately (on different days) from that of precision. If there is no significant time trend in the monitor's mean response (plots or a statistical test such as analysis of variance can provide this information, which could be either within-day or over days), then the intraday relative standard deviation S_{rTE} and interday relative standard deviation S_{rTD} can be estimated by

$$\hat{S}_{rTE} = \sqrt{\mathrm{MSE}(x)}$$

and

$$\hat{S}_{\text{rTD}} = \sqrt{\frac{\text{MSB}(x) - \text{MSE}(x)}{n}},$$

where

$$\text{MSE}(x) = \frac{1}{d(n-1)} \sum_{i=1}^{d} \sum_{j=1}^{n} \left(x_{ij} - \bar{x}_{i\cdot}\right)^2,$$

$$\bar{x}_{i\cdot} = \frac{1}{n} \sum_{j=1}^{n} x_{ij}$$

(B11)

[AppB22], and

$$\text{MSB}(x) = \frac{1}{d-1} \sum_{i=1}^{d} n\left(\bar{x}_{i\cdot} - \bar{x}_{\cdot\cdot}\right)^2,$$

$$\bar{x}_{\cdot\cdot} = \frac{1}{d} \sum_{i=1}^{d} \bar{x}_{i\cdot}.$$

(B12)

[AppB23]. The intramonitor S_{rTW} is then given by

$$\hat{S}_{\text{rTW}} = \sqrt{\hat{S}_{\text{rTE}}^2 + \hat{S}_{\text{rTD}}^2}$$

$$= \sqrt{\left(1 - \frac{1}{n}\right) \times \text{MSE}(x) + \frac{1}{n} \times \text{MSB}(x)}$$

(B13)

[AppB24]. Using the Welch-Satterthwaite formula, the number of degrees of freedom associated with this estimate is

$$v_1 = \frac{\left[\left(1 - \frac{1}{n}\right) \times \text{MSE}(x) + \frac{1}{n} \times \text{MSB}(x)\right]^2}{\dfrac{\left(1 - \frac{1}{n}\right) \times \left[\text{MSE}(x)\right]^2}{dn} + \dfrac{\left[\text{MSB}(x)\right]^2}{(d-1)n^2}}$$

(B14)

[AppB25].

If $\text{MSB}(x) < \text{MSE}(x)$, then $\hat{S}_{\text{rTW}} = \hat{S}_{\text{rTE}}$ and $v_1 = d(n-1)$. It is sensible to check for possible outliers as a cause of this result.

If bias and S_{rT} are independent of concentration (see Section 2.1 in this appendix), then test concentrations can be at any level within the range covered by the bias study. The design presented in this section does not require that the evaluation over days be done at several concentrations.

4.2. Concentration unknown or cannot be reliably replicated

In Section 2.4 of this appendix, it was stated that a pooled estimate of the within-day variance of the test method can be obtained if the F-test described there is not statistically significant. To obtain an estimate of the day-to-day variation, the following procedure can be used. The approach is based on the design discussed in Sections 2.4 and 3.2. The most straightforward way to obtain the day-to-day variation is to evaluate the test unit without the independent method over several days, since once the bias issue is determined, the independent unit is not needed any more. However, the independent method can be used to monitor concentration variability over days. The evaluation is to be done at a single concentration, any concentration in the range under consideration for which the S_{rT} has been determined to be homogeneous, as determined by Section 3.3. For statistical independence, we recommend that the evaluation of bias be done separately (on different days) from that of precision. Insofar as concentration is uncontrollable over days, the between-day variability may be overestimated, although monitoring by the independent method may alleviate this problem.

For the natural logarithms of the unnormalized values (z_{ij}) for the test monitor, where i denotes the ith day $(i = 1, \ldots, d)$ and j denotes the jth test $(j = 1, \ldots, n)$, let $\text{MSE}(z)$ and $\text{MSB}(z)$ denote the mean squares that correspond to (B11) and (B12), but using z_{ij}, instead of x_{ij}. With the above changes

Components for Evaluation of Direct Reading Monitors

in symbols, the estimated S_{rD} for day-to-day variation is given by

$$\hat{S}_{rD}^2 = \left[\mathrm{MSB}(z) - \mathrm{MSE}(z) \right] / n,$$

where the approximation is used that the variance on the natural log scale is approximately the squared relative standard deviation of the original scale data. The degrees of freedom ν_D may be determined by Welch-Satterthwaite's formula:

$$\nu_D = \frac{\hat{S}_{rD}^4}{\dfrac{\left[(1/n)\mathrm{MSE}(z)\right]^2}{d(n-1)} + \dfrac{\left[(1/n)\mathrm{MSB}(z)\right]^2}{d-1}}$$

[AppB26]. The total relative variance can be estimated by

$$\hat{S}_{rW}^2 = \hat{S}_{rD}^2 + \hat{\sigma}_{test}^2.$$

The subscript adds "W" to indicate that it denotes variability "within" a monitor unit, that is, of a single monitor unit, rather than between monitor units. The expression for $\hat{\sigma}_{test}^2$ is given in Section 2.4. By again applying the Welch-Satterthwaite formula, the degrees of freedom of the total variance can be obtained:

$$\frac{\left(\hat{S}_{rD}^2 + \hat{\sigma}_{test}^2 \right)^2}{\dfrac{\hat{S}_{rD}^4}{\nu_D} + \dfrac{\hat{\sigma}_{test}^4}{km-1}}$$

[AppB27]. The estimated total variance relative to the test method mean is transformed to estimated S_{rTW} by $\hat{S}_{rTW}^2 = (1+\hat{B})^2 \hat{S}_{rW}^2$. In the above formula, recall that k is the number of trials and m is the number of independent method determinations in each trial. If the alternative expression for the within day test method relative standard deviation

is used (Section 3.3), then the total relative variance is $\hat{S}_{rD}^2 + \hat{\sigma}_{test,alt}^2$ and the degrees of freedom is:

$$\frac{\left(\hat{S}_{rD}^2 + \hat{\sigma}_{test,alt}^2 \right)^2}{\dfrac{\hat{S}_{rD}^4}{\nu_D} + \dfrac{\hat{\sigma}_{test,alt}^4}{\nu_{test,alt}}}$$

[AppB28].

If $\mathrm{MSB}(z) < \mathrm{MSE}(z)$, then total relative variance can be estimated by $\hat{\sigma}_{test}^2$ or $\hat{\sigma}_{test,alt}^2$, consistent with whether the pooled or alternative expression for the within-day variance is used, with degrees of freedom associated with those estimates.

4.3. Reassessing poolability of S_{rT} values after completion of studies of day-to-day variation

In a laboratory evaluation, large day-to-day variation might not be expected. When the day-to-day variation is large, some explanation should be sought. Also, in this case, it is best to redo the day-to-day evaluation and obtain S_{rTW} estimates for several concentration levels. If there is no way to reduce the day-to-day variation, then the methods provided in Section 3.2.2 can be used to determine groups of poolable S_{rTW} values. A strategy for this decision is given in Section 3 of this appendix.

5. Accuracy Calculations

The S_{rT} obtained by the methods described here can be used to obtain the accuracy confidence limits presented in Appendix A. If there is one component of variance, Equations (A3) and (A4), or (A5) and (A6) of Appendix A can be used, with the choice depending on whether there is not or is bias correction. If there are several components of the test method variance, then use equations (A3) and (A4).

Although the accuracy calculations described here have been based on determination of bias and precision by concentration level and time, there are other factors that can affect both bias and precision. These factors may include temperature, humidity, and the presence of interfering compounds. Usually, in method development, ruggedness testing is carried out to determine that the method is not sensitive to the environmental factors mentioned above. (See the sections, "Environmental Effects" and "Interferences," in Part III, Performance Characteristics.) It is possible to extend the methodology presented here to include the effect of environmental factors on bias and precision.

6. A Strategy for Obtaining Accuracy Estimates That Take into Account Varying Bias or S_{rT} by Concentration Level

Figure B–1 below provides a sequence of decisions that can be used for presenting accuracy estimates based upon precision and bias. The decisions concern the homogeneity of bias (Section 2.2 in this appendix) and of S_{rT} (Section 3 in this appendix). The decisions to be made concern the groupings of concentration levels that will yield both homogeneous S_{rT} and acceptable bias (estimated bias within ±10% of the true value). Although acceptable bias does not require homogeneity of bias over concentration levels, homogeneity is a useful criterion, since only one bias estimate needs to be computed for comparison to the ±10% limit. Procedures were given in Section 2.2.2 of this appendix for determining groups of homogeneous bias; procedures were given in Section 3.2 of this appendix for determining groups of homogeneous S_{rT}. The recommendation is to correct for unacceptable bias by dividing future measurements by (1 + estimated bias), thereby approximately eliminating the bias. However, the method S_{rT} must be adjusted to take into account the variance associated with the bias

correction, as is shown in equations (A5) and (A6) of Appendix A.

The first decision to be made concerns homogeneity of S_{rT} over concentration levels, which determines whether weighted or unweighted least squares should be used in the test of homogeneity of bias, from which acceptability of bias can be determined when true concentrations are known. The test results of homogeneous and acceptable bias and of S_{rT} homogeneity can be used to decide how to present accuracy. If S_{rT} is homogeneous and bias is homogeneous and acceptable, then accuracy computations are done using the pooled values (Option I in Figure B–1). If S_{rT} is homogeneous and bias is not homogeneous, then groupings of poolable bias should be determined, using methods described in Section 2.2.2. Accuracy computations are done for each such group if pooled bias is acceptable (Option II in Figure B–1). If S_{rT} is not homogeneous but bias is, then one of the strategies discussed in Section 3.2 can be used to determine concentration groupings for which accuracy can be determined when bias is acceptable (Option III in Figure B–1). The most complicated situation is when bias is not homogeneous and S_{rT} is not homogeneous. In this case, the largest concentration groupings for which bias is homogeneous and S_{rT} is homogeneous can be determined, and accuracy estimates can be provided for groups with acceptable bias (Option IV in Figure B–1).

Users of the above procedure may wish to allow exceptions based on their professional judgment. For instance, it may be that because of a statistically significant trend in bias, the concentrations with acceptable bias do not have homogeneous bias. There can be situations in which the experimenter nevertheless regards the bias differences as unimportant, in terms of applications, and treats the bias as homogeneous.

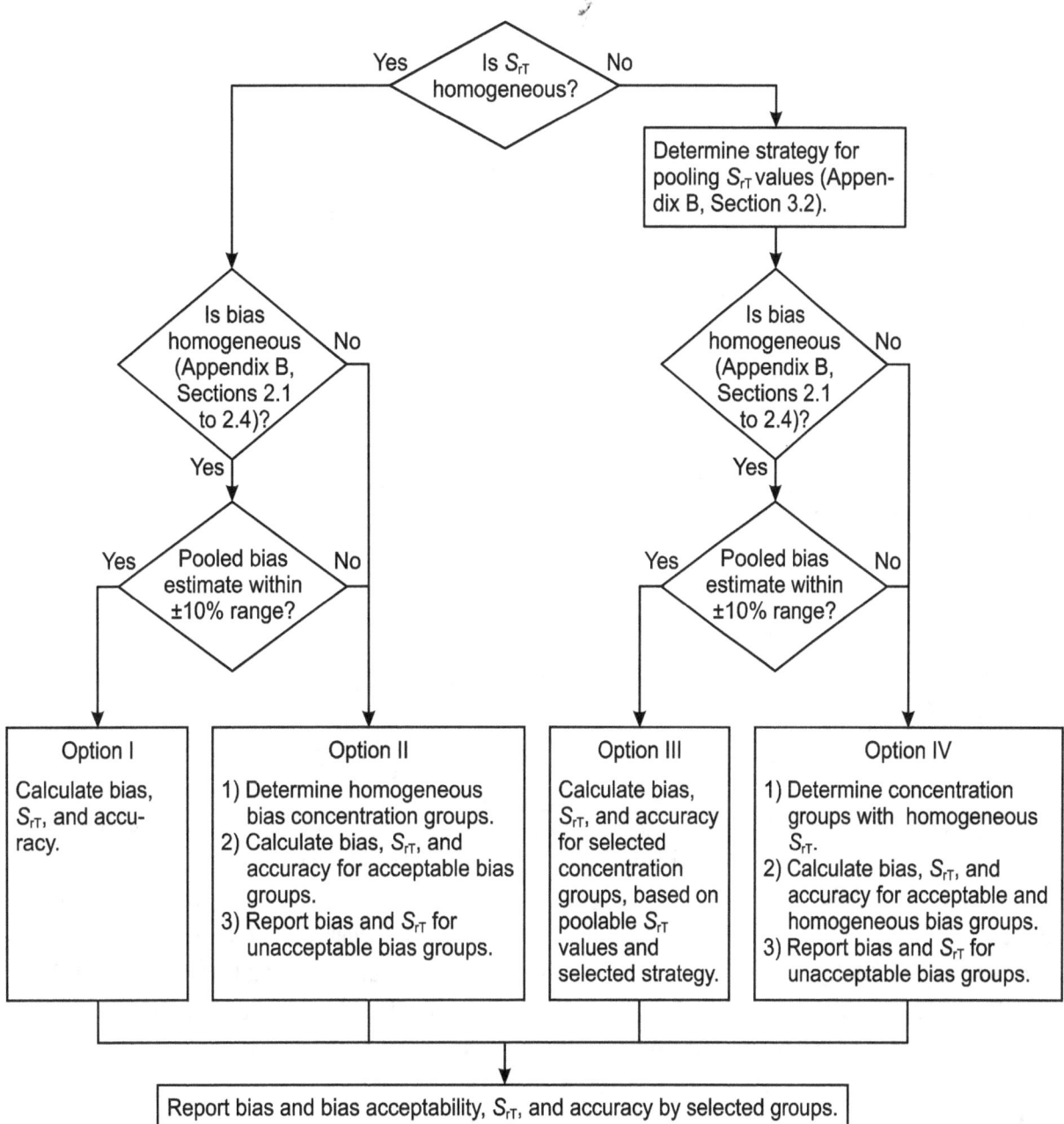

Figure B–1. Decision tree for accuracy estimates.

7. R Code [R Project 2011] for Section 2.4

```
# Test of bias when true concentration cannot be controlled
# k, number of sets
k <- 5
sets <- 1:k
# test monitor determinations, on natural log scale
xt <- c(-0.227, -0.604, 0.974, 2.639, -0.271)
wt <- rep(1, k)
# method = 1 denotes test method
```

```
method <- rep(1, k)
xt <- cbind(sets, method, xt, wt)
# independent method determination means, from data on natural log scale
xtn <- c(-0.355, -0.551, 0.342, 2.390, -1.257)
# independent method sample variances, from data on natural log scale
var <- c(0.0207, 0.260, 0.0721, 0.0574, 0.0448)
# m, number of independent method determinations in each set
m <- 3
wt <- rep(m, k)
# method = 2 denotes independent method
method <- rep(2, k)
xi <- cbind(sets, method, xtn, wt)
xx <- rbind(xt, xi)
xx_fr <- data.frame(xx)
xx_fr$sets <- as.factor(xx_fr$sets)
xx_fr$method <- as.factor(xx_fr$method)
lm1 <- lm(xx_fr$xt ~ xx_fr$sets + xx_fr$method, weights = xx_fr$wt)
# sig2_bias is the square of eq (B9b)
sig2_bias <- (sum(lm1$residual^2 * xx_fr$wt)) / (k - 1)
# sig2_ind is the square of sigma_res
sig2_ind <- mean(var)
# G statistic from section 2.4
G <- sig2_bias / sig2_ind
# p-value of G statistic; if less than 0.05, reject homogeneity of bias
1 - pf(G, k - 1, k * (m - 1))
```

8. References

Barnett V, Lewis T [1978]. Outliers in statistics. New York: John Wiley & Sons, pp. 93–94.

Bartlett MS [1937]. Some examples of statistical methods of research in agriculture and applied biology. J Royal Stat Soc 4(Suppl):58–159.

Hald A [1952]. Statistical theory with engineering applications. New York: John Wiley & Sons, p. 291.

Miller R Jr. [1981a]. Simultaneous statistical inference. 2nd ed. New York: Springer-Verlag, pp. 37–48.

Miller R Jr. [1981b]. Simultaneous statistical inference. 2nd ed. New York: Springer-Verlag, p. 222.

R Project [2011]. The R project for statistical computing [http://www.r-project.org/]. Date accessed: November 2011.

Scheffe H [1959]. Analysis of variance. New York: John Wiley & Sons, section 3.8.

Song R, Kennedy ER, Bartley DL [2001]. Uniformity test of bias when reference value contains experimental error. Anal Chem 73(2):310–314.

Welch BL [1956]. On linear combinations of several variances. J Am Stat Assoc 51:132–148.

9. Related Sources

NIOSH [1995]. Guidelines for air sampling and analytical method development and evaluation. By Kennedy ER, Fischbach TJ, Song R, Eller PM, Shulman SA. Cincinnati, OH: U.S. Department of Health and Human Services, Centers for Disease Control and Prevention, National Institute for Occupational Safety and Health, DHHS (NIOSH) Publication No. 95–117. [http://www.cdc.gov/niosh/docs/95-117/pdfs/95-117.pdf]

Appendix C. Statistical Evaluation of Bias and Precision for a Population of Monitor Units

1. Introduction

The methods presented in Appendix B were provided for a single monitor unit. There are situations where results are required for multiple units:

(1) If the monitor being used has not been evaluated, then the required information for this monitor can be obtained from other monitors of the same brand. This would be expected to be a conservative procedure, since intermonitor variability would be expected to exceed intramonitor variability.

(2) The manufacturer may be interested in determining the accuracy of the population of monitor units of the same kind.

Computations are more complicated for the case of multiple monitor units. For accuracy evaluation, both bias and precision are required, as for a single monitor. A variety of experimental designs is possible. Perhaps the simplest design would evaluate the monitors at each of several concentration levels. This would make possible an assessment of bias and intermonitor variability. If the evaluation can be repeated over several days, then day-to-day variability can also be estimated.

2. Estimation of Bias

The types of data are the same as those discussed in Appendix B.

2.1. True concentrations known or stoichiometrically calculated target used

Divide monitor determinations by the true concentration or target to obtain the normalized test result y_{ijl}, where i denotes the monitor unit ($i = 1,\ldots,u$), j denotes the day ($j = 1,\ldots,d$), and l denotes the trial in day j ($l = 1,\ldots,k$). For bias, the data could be collected for just one day, although the design below

presents results based on several days of data. Assume that there are u units, d days, and k trials on each day.

Mean squares $MS_U(y)$, $MS_D(y)$, $MS_{UD}(y)$, $MS_{tD}(y)$, and $MS_{res}(y)$, are defined as follows:

$$MS_U(y) = \frac{kd\sum_{i=1}^{u}\left(\bar{y}_{i..} - \bar{y}_{...}\right)^2}{u-1}, \quad (C1a)$$

$$\bar{y}_{i..} = \sum_{j,l} y_{ijl}/(kd),$$

$$\bar{y}_{...} = \sum_{i,j,l} y_{ijl}/(kdu)$$

[AppC01],

$$MS_D(y) = \frac{ku\sum_{j=1}^{d}\left(\bar{y}_{.j.} - \bar{y}_{...}\right)^2}{d-1}, \quad (C1b)$$

$$\bar{y}_{.j.} = \sum_{i,l} y_{il}/(ku)$$

[AppC02],

$$MS_{UD}(y) = \frac{k\sum_{i,j}\left(\bar{y}_{ij.} - \bar{y}_{i..} - \bar{y}_{.j.} + \bar{y}_{...}\right)^2}{(u-1)(d-1)}, \quad (C1c)$$

$$\bar{y}_{ij.} = \sum_{l=1}^{k} y_{ijl}/k$$

[AppC03],

$$MS_{tD}(y) = \frac{\left[\begin{array}{c} u\sum_{j,l}\left(\bar{y}_{.jl} - \bar{y}_{...}\right)^2 \\ -(d-1)MS_D(y) \end{array}\right]}{d(k-1)}, \quad (C1d)$$

$$\bar{y}_{.jl} = \sum_{i=1}^{u} y_{ijl}/u$$

[AppC04],

$$\text{MS}_{\text{res}}(y) = \frac{\begin{bmatrix} \sum\limits_{i,j,l} \bar{y}_{ijl}^2 - (kdu)\bar{y}_{...}^2 \\ -(u-1)\text{MS}_{\text{U}}(y) \\ -(d-1)\text{MS}_{\text{D}}(y) \\ -(u-1)(d-1)\text{MS}_{\text{UD}}(y) \\ -d(k-1)\text{MS}_{\text{tD}}(y) \end{bmatrix}}{d(u-1)(k-1)} \quad \text{(C1e)}$$

[AppC05]. In the above expressions, when the limits of summation are not shown, the index runs from its lowest to highest value. For example,

$$\sum_{i,l} \equiv \sum_{i=1}^{u} \sum_{l=1}^{k}$$

[AppC06].

Under normality assumptions, the ratio $\text{MS}_{\text{tD}}(y)/\text{MS}_{\text{res}}(y)$ is distributed as an F-statistic with $d(k-1)$ and $d(u-1)(k-1)$ degrees of freedom. If the F-statistic does not yield a statistically significant result then the bias can be said to be homogeneous over concentration levels. The above mean squares can be obtained from any statistical program that does analysis of variance. This test is presented under the assumption that the same test concentrations may not be repeated for all days, which is probably the most general approach. However, the approach will work best if approximately the same concentrations are used on each day. Also, the test must allow for sufficient spread in the test concentrations in order to evaluate homogeneity of bias. Note that if there were only data for one day, the only mean squares would be $\text{MS}_{\text{U}}(y)$, $\text{MS}_{\text{tD}}(y)$, and $\text{MS}_{\text{res}}(y)$.

The above design will reject homogeneity of bias if there is large variability of the bias by concentration over days. For this kind of data, where the true concentration is known or can be targeted by stoichiometric calculations, it should be possible to repeat the same trial concentrations on each day. In this case, the $\text{MS}_{\text{tD}}(y)$ can be separated into two mean squares, one for trial concentrations and one for trial concentrations by days, and the factors corresponding to these mean squares can be evaluated separately. Statistical significance for either of these mean squares relative to the appropriate error terms is reason to decide that the test method biases are not homogeneous over concentration levels. Appropriate testing may require consultation with a statistician. In addition, for this modification of the design the estimation of a mean square for units by trial concentration makes sense, the statistical significance of which can be tested by comparison to the mean square residual. This comparison, too, is part of the evaluation of the homogeneity of bias.

If the various tests suggested above lead to statistically significant results, then it may be advisable to divide the data into groups based on test concentration and redo the tests for these subgroups. The analysis could stop when groupings are determined for which the tests do not yield statistically significant results. However, there is judgment involved as to which level of significance should be of concern, since there will always be some variability of measurements by concentration. For this reason, the analysis of these results should be done in consultation with a statistician.

Although the aim of these analyses is the assessment of homogeneity of bias by trial concentration, the proposed tests also make assessments of homogeneity of precision. To conclude that bias does not vary by trial concentrations over days is to say that the associated variance is small. Other variance components, for days or for units over days, will contribute to the variance of y_{ijl}. Recall that the variance of y_{ijl} is the S_{rT}^2 of the unnormalized measurement.

2.2. True concentration not known

The data may be viewed as arising from the format presented in Table C–1 below, in which there are u determinations by independent and test methods, all made at the same time (or under identical conditions) on each of k trials during d days. For the test method, the "u determinations" refer to determinations by u monitors, the same monitors to be used in each trial in the study. For independent method samples, the meaning of "u determinations" depends on the independent method: 1) all samples may be statistically independent samples, or 2) samples may be determinations by the same u (>1) monitor units in each trial, determinations by the same monitor being statistically dependent. If the independent method uses monitors, but just one monitor unit is evaluated, then there is no way to account for intermonitor variability, and situation (1) will be assumed. Also, there is no reason that each method must have the same number (u) of determinations; these could differ. This is only done for simplicity of presentation. Note that the x_{ijl} values in the table are the actual measurements. The test and independent method samplers are assumed to be placed randomly. Alternatively (though not considered here), there could be pairings of test and independent method samplers, useful if there is substantial uncontrolled variability.

Table C–1. Experimental design for comparison of test and independent methods

Day	Trial	Test method		Independent method	
		Determination	**Result**	**Determination**	**Result**
		1	$x_{111(\text{test})}$	1	$x_{111(\text{ind})}$
	1	2	$x_{211(\text{test})}$	2	$x_{211(\text{ind})}$
		$\vdots$	$\vdots$	$\vdots$	$\vdots$
		u	$x_{u11(\text{test})}$	u	$x_{u11(\text{ind})}$
1	$\vdots$	$\vdots$	$\vdots$	$\vdots$	$\vdots$
		1	$x_{11k(\text{test})}$	1	$x_{11k(\text{ind})}$
	k	2	$x_{21k(\text{test})}$	2	$x_{21k(\text{ind})}$
		$\vdots$	$\vdots$	$\vdots$	$\vdots$
		u	$x_{u1k(\text{test})}$	u	$x_{u1k(\text{ind})}$
$\vdots$	$\vdots$	$\vdots$	$\vdots$	$\vdots$	$\vdots$
		1	$x_{1d1(\text{test})}$	1	$x_{1d1(\text{ind})}$
	1	2	$x_{2d1(\text{test})}$	2	$x_{2d1(\text{ind})}$
		$\vdots$	$\vdots$	$\vdots$	$\vdots$
		u	$x_{ud1(\text{test})}$	u	$x_{ud1(\text{ind})}$
d	$\vdots$	$\vdots$	$\vdots$	$\vdots$	$\vdots$
		1	$x_{1dk(\text{test})}$	1	$x_{1dk(\text{ind})}$
	k	2	$x_{2dk(\text{test})}$	2	$x_{2dk(\text{ind})}$
		$\vdots$	$\vdots$	$\vdots$	$\vdots$
		u	$x_{udk(\text{test})}$	u	$x_{udk(\text{ind})}$

The test must allow for sufficient spread in the trial concentrations in order to evaluate homogeneity of bias. It is sensible to attempt to have approximately the same minimum and maximum concentration for each day's trials. Although this is not necessary for the statistical method presented here, it would allow for better estimates of the day-to-day and within-day variation, and these may be of interest, in addition to the total method variation.

Let: $z_{ijl(\text{test})}$ = ln (monitor measurement) for the test unit data

$$= \ln(x_{ijl(\text{test})})$$

and

$z_{ijl(\text{ind})}$ = ln (determination) for independent method data

$$= \ln(x_{ijl(\text{ind})}).$$

When the independent method uses independent samples for each determination, the analysis involves a somewhat complicated mixed model that can be fitted with the help of a statistician. For the simpler case that the independent method uses monitor units, the following discussion applies.

For the purposes of the bias analysis, we will add an additional subscript g, so that $z_{1ijl} = z_{ijl(\text{test})}$ and $z_{2ijl} = z_{ijl(\text{ind})}$. Using this notation we can define the mean squares for methods $\text{MS}_m(z)$, for unit within methods $\text{MS}_{\text{Um}}(z)$, for days $\text{MS}_D(z)$, for trials in days $\text{MS}_{tD}(z)$, for methods by days $\text{MS}_{mD}(z)$, for units in methods by days $\text{MS}_{mUD}(z)$, for methods by trials in days $\text{MS}_{mtD}(z)$, and for the residual $\text{MS}_{\text{res}}(z)$. These mean squares can be produced by any statistical package that has an analysis of variance program. Under normality assumptions, the ratio $\text{MS}_{mtD}(z)/\text{MS}_{\text{res}}(z)$ follows an F-distribution with $d(k-1)$ and $2d(k-1)(u-1)$ degrees of freedom. If the 95th percentile of that distribution is exceeded, that is reason to think that the method bias varies by concentration, since the different trials are at different concentrations.

As was discussed in Section 2.1, if it is possible to repeat the same concentration levels on each day of the evaluation, then the $\text{MS}_{mtD}(z)$ can be separated into mean squares for method by trial concentrations and for method by trial concentrations over days. Separate statistical analyses would then be possible to assess the significance of method differences by test concentrations over days and of average differences of methods over test concentrations. Statistical significance for either of the associated mean squares relative to the appropriate error terms is reason to decide that the method biases are not homogeneous over concentration levels. As in Section 2.1, for this modification of the design the estimation of a mean square for differences of methods by units by trial concentration makes sense, the statistical significance of which can be tested by comparison to the mean square residual. This comparison, too, is part of the evaluation of the homogeneity of bias.

For the kind of data where there is no reliable way to replicate the concentration, the procedure described above [evaluation based on $\text{MS}_{mtD}(z)/\text{MS}_{\text{res}}(z)$] should be adequate to assess either significant differences of methods by concentration or by concentration over days.

As was stated in Section 2.1, if the various tests suggested above lead to statistically significant results, then it may be advisable to divide the data into groups based on test concentration and redo the tests for these subgroups. The analysis could stop when groupings are determined for which the tests do not yield statistically significant results. However, there is judgment involved as to which level of significance should be of concern, since there will always be some variability of measurements by concentration. For this reason, the analysis of these results should be done in consultation with a statistician.

The pooled estimate of bias $\hat{B}$ is:

$$\hat{B} = \exp\left(\overline{z}_{\cdots(\text{test})} - \overline{z}_{\cdots(\text{ind})}\right)$$

[AppC07]. As with the evaluation of single units, if the bias is not homogeneous, then groups of concentrations should be determined in which there is homogeneity.

3. Estimation of Precision

There are four nonconcentration components of variability that contribute to the total measurement variance when there is homogeneity of precision over concentration levels: variability between monitor units, variability of units within days, variability of units between days, and variability between days. Conversely, inhomogeneity of precision over concentration levels means that the variability of measurements over concentrations does vary appreciably and does vary appreciably by unit or by day. As described in Section 2, subgroups are to be formed so that this variability is small. For the design where the concentration cannot be reliably replicated, this assumption may be difficult to verify. Even when subgroups are formed, it is unlikely that the variances associated with different concentrations will be zero.

3.1. True concentrations known or stoichiometrically calculated target known

3.1.1. Day-to-day variation is based on the evaluation of a single monitor

To obtain a total S_{rT}, an evaluation of at least one of the monitors for variation over time, as in Appendix B, Section 4.1, must also be done. To obtain the intermonitor S_{rTB}, a number of monitors of the same brand and model number need to be tested simultaneously. Several trials are necessary to get a better estimate for the intermonitor S_{rTB}. If possible, these trials are done within one day to eliminate the day-to-day variability from this analysis. If bias or S_{rT} could depend on concentration, then a different concentration can be used for each

trial. Because there is no day-to-day variation in the intermonitor evaluation, the notation of Section 2.1 is altered as follows. Let y_{il} be the normalized test result of the ith monitor unit, $i = 1, \ldots, u$, at time t_l, $l = 1, \ldots, k$, where these times are all on the same day. Then, after allowing for individual means for each time t_l, the intermonitor relative standard deviations can be estimated by

$$\hat{S}_{rTB} = \sqrt{\frac{MSB(y) - MSE(y)}{k}}$$

[AppC08], where

$$\bar{y}_{i.} = \frac{1}{k}\sum_{l=1}^{k} y_{il}, \bar{y}_{.l} = \frac{1}{u}\sum_{i=1}^{u} y_{il}, \bar{y}_{..} = \frac{1}{u}\sum_{i=1}^{u} \bar{y}_{i.}$$

[AppC09],

$$MSE(y) = 1/\left[(k-1)(u-1)\right]$$
$$\times \sum_{i=1}^{u}\sum_{l=1}^{k}(y_{il} - \bar{y}_{i.} - \bar{y}_{.l} + \bar{y}_{..})^2$$

[AppC10], and

$$MSB(y) = \frac{1}{u-1}\sum_{i=1}^{u} k(\bar{y}_{i.} - \bar{y}_{..})^2 \qquad (C2)$$

[AppC11]. In (C2), $MSB(y)$ corresponds to $MS_U(y)$ of (C1a). The number of degrees of freedom associated with $\hat{S}_{rTB}$ is

$$v_2 = \frac{\left\{\left[MSB(y) - MSE(y)\right]/k\right\}^2}{\dfrac{\left[(1/k)MSB(y)\right]^2}{u-1} + \dfrac{\left[(1/k)MSE(y)\right]^2}{(u-1)(k-1)}}$$

[AppC12].

In this case, the total $\hat{S}_{rT}$ for accuracy evaluation must include both the intra- and inter-monitor $\hat{S}_{rT}$ values:

$$\hat{S}_{rT} = \sqrt{\hat{S}_{rTW}^2 + \hat{S}_{rTB}^2}$$

[AppC13]. The number of degrees of freedom associated with this total $\hat{S}_{rT}$ is:

$$v = \frac{\left[\hat{S}_{rTW}^2 + \hat{S}_{rTB}^2\right]^2}{\hat{S}_{rTW}^4/v_1 + \hat{S}_{rTB}^4/v_2}.$$

$\hat{S}_{rTW}$ and v_1 were given in (B13) and (B14) of Appendix B.

Homogeneity of bias can also be tested by a test analogous to equation (B6) of Appendix B, in which the numerator is analogous to that in equation (B5) of Appendix B, and the denominator is MSE(y). If the test produces a statistically significant result, then groupings of concentration levels should be made with approximately constant bias within each group.

3.1.2. Within-unit and between-unit precision are both estimated using all monitor units

In general, the estimated S_{rT} will include several of the variance components associated with the mean squares shown in (C1). Because the formula for the total variance becomes quite complicated when many variances are used, it is not sensible to provide a general formula. A statistician should be consulted.

It seems useful to provide an example of the form of the estimated S_{rT} for the case that the independent method evaluation is based on u monitor units. Suppose that the variance components associated with concentration are all negligible, and suppose that the day-to-day variance is also small. The $MS_{res}(y)$ in (C1e) provides an estimate of S_{rTW} with $d(u-1)(k-1)$ degrees of freedom. As in MSB(y) in (C2) of Section 3.1.1, $MS_U(y)$ in (C1a) provides an estimate of the between monitor variability, though for the multiday evaluation, it must be combined with $MS_{UD}(y)$ in (C1c) as follows:

$$\hat{S}_{rT}^2 = \frac{k-1}{k}MS_{res}(y) + \frac{d-1}{dk}MS_{UD}(y) + \frac{1}{dk}MS_U(y)$$

[AppC14]. The Welch-Satterthwaite approximation to the degrees of freedom [Welch 1956] of $\hat{S}_{rT}$ is:

$$\frac{\hat{S}_{rT}^4}{\left(\dfrac{\left[(1-1/k)MS_{res}(y)\right]^2}{d(k-1)(u-1)} + \dfrac{\left[(1/k-1/(dk))MS_{UD}(y)\right]^2}{(d-1)(u-1)} + \dfrac{\left[(1/(dk))MS_U(y)\right]^2}{(u-1)}\right)} \tag{C3}$$

[AppC15].

The above formulas are based on the mean squares from the statistical model used in (C1). If, as assumed, the day-to-day variance and the variance components associated with concentration are all small, then they can be pooled with the residual, and the above formula would change. A statistician should be consulted.

3.2. True concentration not known

For estimation of precision, the natural log scale of the unnormalized test unit measurement $z_{ijl(\text{test})}$ is used as the dependent variable; standard deviations on the natural log scale are approximately equal to S_r values on the original scale. The analysis of variance can be used to obtain estimates of the variance components, which would be summed to obtain a total relative variance estimate. When there is no reliable way to replicate concentration, it would be sensible to check from the residual plots that the within-day variance does not depend on concentration. The degrees of freedom for the estimated total variance can be obtained by a version of Welch-Satterthwaite's formula. The formula, $S_{rT} = S_r \times (1 + B)$, is needed to obtain the total S_{rT} values for the accuracy analysis, although for inhomogeneous bias, the appropriate value for B would vary for the different homogeneous groups. Appropriate software is available in R [R Project 2011].

In general, the estimated S_{rT} will include several of the variance components. Because the formula for the total variance becomes quite complicated when many variances are used, it is not sensible to provide a general formula. A statistician should be consulted.

3.3. Accuracy calculations

For accuracy calculations, the degrees of freedom of the estimated total variance is required, as given in (C3). If N is chosen to be one greater than the degrees of freedom, then the equations (A3) and (A4) in Appendix A may be used.

4. References

R Project [2011]. The R project for statistical computing [http://www.r-project.org/]. Date accessed: November 2011.

Welch BL [1956]. On linear combinations of several variances. J Am Stat Assoc *51*:132–148.

Appendix D. Measurement Uncertainty

1. Uncertainty Analysis

The international standard, ISO/IEC 17025 [ISO 2005], requires that for each measurement result reported, a measurement uncertainty associated with the result should be provided. To meet this requirement, a monitor should have a complete performance evaluation before it is used in the workplace. At the least, representative units of the same brand monitor should be evaluated. In this case, the intermonitor variability must be included in the final value of measurement uncertainty.

Measurement uncertainty may be expressed in terms of standard deviation or relative standard deviation depending on which one of the two precision parameters is independent of the concentration of the analyte being measured [ISO 1993]. Since the relative standard deviation is more likely to be constant, the measurement uncertainty in the form of relative standard deviation is used here, as it was used in Appendix B.

The measurement uncertainty can be used to construct an uncertainty interval around the measurement result y, such that the resulting interval has a high probability of covering the true concentration C_T. There are two kinds of measurement uncertainty: the standard uncertainty u and the expanded uncertainty U. The standard uncertainty is an estimate of the standard deviation (or relative standard deviation). The expanded uncertainty is a quantity derived from the standard uncertainty having the form $U = k \times u$. The multiplier k, called coverage factor, is selected such that the uncertainty interval

$$\left(\frac{y}{1+ku}, \frac{y}{1-ku} \right)$$

[AppD01] or, equivalently,

$$\left(y - kuC_T, y + kuC_T \right) \qquad \text{(D1)}$$

has a prespecified probability p to cover the true concentration C_T, and u denotes the estimated uncertainty relative to C_T. (Note that k is not used here to represent the number of concentration levels.) If there is no significant bias, or bias has been corrected in the final result y (e.g., by division of raw estimates by $(1+B)$ if the bias is constant), and if u is a standard deviation relative to true concentration S_{rT}, then the desired coverage factor [ISO 1993]

$$k = t_{(1+p)/2,v} \qquad \text{(D2)}$$

[AppD02] is the $\left[(1+p)/2 \right] \times 100$ percentile of the t-distribution, where v is the number of degrees of freedom associated with the standard uncertainty.

2. Comparing Uncertainty and Accuracy

Alternatively, (D1) may be written, more simply, as:

$$y - U < C_T < y + U.$$

If there is no significant bias, or bias has been corrected in the final result y (e.g., by division of raw estimates by $(1+B)$ if the bias is constant), then the desired coverage factor [ISO 1993]

$$k = t_{(1+p)/2,v}$$

[AppD03] is the $\left[(1+p)/2 \right] \times 100$ percentile of the t-distribution, where v is the number of degrees of freedom associated with the standard uncertainty.

The accuracy A, as defined by (A1) and (A2), satisfies the following relationship:

$$C_T \times (1-A) < y < C_T \times (1+A) \qquad \text{(D3)}$$

for 95% of measurements y. Since (D3), if A is much smaller than 1, can be rewritten as

$$y - (y \times A) < C_T < y + (y \times A), \qquad \text{(D4)}$$

which applies with 95% confidence, therefore, $U = y \times A$ is the expanded uncertainty for the 95% confidence statement. Thus, for small A there is approximate equivalence of accuracy to uncertainty.

Suppose bias is known to be negligible. Then:

$$\begin{aligned} A &= 1.960 \times S_{rT} \\ &= 1.960 \times \sqrt{v/\chi^2_{0.05,v}} \times \hat{S}_{rT} \end{aligned} \qquad \text{(D5)}$$

[AppD04], where $\chi^2_{0.05,v}$ [AppD05] is the 5th percentile of the chi square distribution with v degrees of freedom.

For $v = 15$, the coverage factor k is:

$$\begin{aligned} k &= 1.960 \times \sqrt{15/\chi^2_{0.05,15}} \\ &= 2.8 \end{aligned}$$

[AppD06]. For comparison, equation (D2), with $p = 0.95$ and 15 degrees of freedom, yields a k value of 2.13. On the other hand, it is suggested in uncertainty references that k be chosen as 2 or 3, and the calculated k of 2.8 falls in between.

Although there is comparability between uncertainty and accuracy intervals, there are some differences. Perhaps the easiest way to understand the difference in interpretation is to recognize the use of the chi square distribution in (D5). Equation (A2) provides a definition for accuracy but not a means of estimating it. Estimation requires data, and the value 0.05 is the 5th percentile for an experiment to estimate the standard deviation of the measurements. The accuracy calculation is based on the evaluation experiment. Appendix B discusses various kinds of evaluation experiments. As presented in this document, the accuracy experiment is intended to provide the chances of attaining a certain accuracy (usually 95% of the population should be within 25% of the true value) with 95% confidence, for future uses of the method, under conditions similar to those used in the initial evaluation.

By contrast, uncertainty is intended to apply to a particular measurement. Since the aim is to identify all sources of variability, and since uncertainty procedures attempt to correct for all biases, the expression (D1) includes only uncertainty, not bias. Since the components of uncertainty are specified by the developer of the estimate of uncertainty (just as the choice of the method evaluation experiment used in accuracy estimation is due to the experimenter), it is possible that the expanded uncertainty can be greater than or less than the $(y \times A)$ values that appear in (D4).

There is no reason that uncertainty evaluation cannot be made part of the method evaluation process. The example experiments shown in Table 1 (in Part III, Monitor Evaluation Data Reduction) could be part of that process. In many ways, the bias and precision experiments discussed in Appendices B and C are evaluations of whether the biases have been removed and whether the total uncertainty has been approximated.

3. References

ISO [1993]. Guide to the expression of uncertainty in measurement. Geneva, Switzerland: International Organization for Standardization.

ISO [2005]. ISO/IEC 17025 General requirements for the competence of testing and calibration laboratories. Geneva, Switzerland: International Organization for Standardization.

Appendix E. Relationship of the NIOSH Accuracy Criterion to Monitor Performance Specifications

In these *Components*, the performance of a monitor is often based on a requirement that the monitor must provide a response under a given set of conditions of ±Z% of the true concentration. To satisfy the NIOSH accuracy criterion (AC) (95% of the measurement are within ±25% of the true concentration with 95% confidence), m out of a total of n measurement results for the monitor must be within ±Z% of the true concentration. The number n is prespecified, and m is determined, based on the NIOSH AC.

If a monitor meets the NIOSH AC, then this monitor should have a probability of 56.7% or greater for results to fall within ±10% of the true concentration, if measurement results are normally distributed. To make sure that the monitor meets this criterion with at least 95% confidence, a lower confidence limit estimate for the fraction (m of n) of results that meets the criterion is constructed. For probability ≥ 56.7% and selected n, the values for m are listed in Table E–1 below.

The underlying assumptions for the application of this approach are: (1) measurement results are normally distributed and (2) the method under study is unbiased. Under these assumptions, the NIOSH AC that 95% of results fall in ±25% from the true concentration (p/q criterion with $p = 95$ and $q = 25$) is statistically equivalent to the 56.7 / 10 criterion because both criteria require that the method has a precision ($S_r = S_{rT}$ since no bias is assumed) less than 12.8%.

The two criteria are different when the second assumption is not true. The 95 / 25 criterion can tolerate a larger bias than the 56.7 / 10 criterion. For example, a method having a bias = 10% and a precision $S_r = 5\%$ meets the 95 / 25 criterion, but not the 56.7 / 10 criterion.

Under the above two assumptions, the NIOSH AC can be expressed in the form that p% of results fall in ±q% from the true concentration, with p and q dependent on each other. When the bias assumption is not met, the quantity q places a limit on bias. In this case, there is no need to set an additional requirement on bias.

Table E–1. Number m of n measurements that must be within 10% of target to meet NIOSH accuracy criterion

n	m (criterion)
6	6
10	9
15	13
20	16
25	19
30	22
50	35
100	66

Appendix F. Alarm System Evaluation

1. Introduction

The details behind evaluating an alarm system so as to control false positives and negatives are presented here. The basic idea is similar to that used in the NIOSH accuracy criterion for evaluating sampling and analytical methods. In this case, the range of a specified fraction of estimates about true concentrations is estimated at confidence γ (specifically 95%) in the evaluation. For an alarm system at confidence γ in the evaluation, cut-offs for the alarm system where the alarm should be **on** or where it should be **off** are estimated so that false positives and negatives are under control. The statistical approach applied involves tolerance intervals—symmetric for NIOSH accuracy, and single-sided for alarm systems.

The details below are somewhat complicated, particularly with the anticipation that many alarm systems will be applied at low concentrations, i.e., near the limit of quantitation, where neither the system's relative bias nor relative standard deviation is independent of the sampled concentration. The statistical analysis is simplified markedly by evaluating the system only at two levels: C_{alarm}, the true concentration where the alarm should be **on**, and C_0, a safe concentration (e.g., 0), where alarm should not be on. However, more information is obtained by evaluation over a range of true concentrations $C_j, j = 1, \ldots, k$ (e.g., 4), encompassing C_{alarm}. In any case, the complications are invisible to the user of programs (as presented in Appendix G) for analyzing the results of the evaluation.

In sampling true concentration C, suppose the alarming instrumentation produces an output $\hat{c}$. Suppose $\hat{c}$ is related to C by

$$\hat{c} = c + \hat{\varepsilon} \tag{F1}$$

[AppF01], where

$$c = \delta_0 + (1 + \delta_1)C \tag{F2}$$

[AppF02] and $\hat{\varepsilon}$ is normally distributed about zero with variance σ^2,

$$\sigma^2 = \sigma_0^2 + \sigma_1^2 C^2 \tag{F3}$$

[AppF03].

The various constants are distinguished as follows: δ_0 is an intercept, δ_1 is usually denoted as the relative bias, σ_0^2 is a variance component responsible for finiteness of LOD, and σ_1^2 is the asymptotic squared S_{rT}, where S_{rT} is the concentration measurement standard deviation relative to the true concentration.

Furthermore, define: α_-, the permissible false negative error rate (e.g., 5%) when $C = C_{\text{alarm}}$, and α_+, the permissible false positive error rate (e.g., 5%) when $C = C_0$.

2. Controlling False Negatives (σ and c Known at $C = C_{\text{alarm}}$)

Let c_{alarm} denote the output above which the system's alarm is **on**. The definition of the false negative rate α_- implies that at $C = C_{\text{alarm}}$,

$$\frac{1}{\sqrt{2\pi}\sigma} \int_{-\infty}^{c_{\text{alarm}}} e^{-\frac{1}{2}(\hat{c}-c)^2/\sigma^2} \, d\hat{c} = \alpha_-$$

[AppF04]. In other words,

$$c_{\text{alarm}} = c[C_{\text{alarm}}] + u_{\alpha_-}\sigma[C_{\text{alarm}}] \tag{F4}$$

[AppF05], where u_{α_-} is the normal quantile at level α_- (e.g., $u_{0.05} = -1.645$).

3. σ and c Unknown

In general, σ and c are unknown. In this case, an evaluation experiment must be carried out to determine a reasonable value c_{alarm} at which the alarm sounds. The basic idea is to conduct a sufficiently extensive experiment in which the functions c and σ^2 [equations (F2) and (F3)] are adequately estimated so that a useful confidence limit on the right-hand side of equation (F4) may be approximated.

Obtain n (e.g., 10) replicate estimates $\hat{c}_j$ at each of k (e.g., 4) values of the true concentration $C_j, j = 1, \ldots, k$. Then s_j^2 values, the (usual) variance estimates with $n-1$ degrees of freedom, are easily computed at each value of j. The variance parameters $\hat{\sigma}_0^2$ and $\hat{\sigma}_1^2$ are then obtained from linear regression as

$$\hat{s}_j^2 = \hat{\sigma}_0^2 + \hat{\sigma}_1^2 C_j^2$$

[AppF06]. Similarly, the replicate averages $\overline{c}_j$ are easily computed and, then, the bias parameters δ_0 and δ_1 may be obtained, also from ordinary linear regression:

$$\overline{c}_j = \hat{\delta}_0 + \left(1 + \hat{\delta}_1\right) C_j$$

[AppF07].

Needed below is the variance of the predictor c_{est}, $c_{est} = \hat{\delta}_0 + (1 + \hat{\delta}_1) C_{alarm}$, at the alarm concentration C_{alarm}:

$$\mathrm{var}\left[c_{est}\right] = \mathrm{var}\left[\langle \overline{c} \rangle + \left(C_{alarm} - \langle C \rangle\right)\hat{\delta}_1\right]$$

[AppF08], where the angle brackets $\langle \ \rangle$ denote averages over $j = 1, \ldots, k$ (henceforth omitting the subscript j). The variance $\mathrm{var}[c_{est}]$ can be evaluated as follows, recognizing that the variance σ_j^2 may not be constant over j:

$$\mathrm{var}\left[\langle \overline{c} \rangle\right] = \frac{1}{kn}\langle \sigma^2 \rangle$$

[AppF09],

$$\mathrm{var}\left[\langle \hat{\delta}_1 \rangle\right] = \frac{1}{kn} \frac{\left\langle \sigma^2 \left(C - \langle C \rangle\right)^2 \right\rangle}{\left\langle \left(C - \langle C \rangle\right)^2 \right\rangle^2}$$

[AppF10], and

$$\mathrm{cov}\left[\langle \overline{c} \rangle, \hat{\delta}_1\right] = \frac{1}{kn} \frac{\left\langle \sigma^2 \left(C - \langle C \rangle\right) \right\rangle}{\left\langle \left(C - \langle C \rangle\right)^2 \right\rangle}$$

[AppF11].

Because

$$\mathrm{var}\left[\langle \overline{c} \rangle + \left(C_{alarm} - \langle C \rangle\right)\left(1 + \hat{\delta}_1\right)\right] =$$
$$\mathrm{var}\left[\langle \overline{c} \rangle\right] + \left(C_{alarm} - \langle C \rangle\right)^2 \mathrm{var}\left[\hat{\delta}_1\right]$$
$$+ 2\left(C_{alarm} - \langle C \rangle\right)\mathrm{cov}\left(\langle \overline{c} \rangle, \hat{\delta}_1\right)$$

[AppF12], therefore,

$$\mathrm{var}\left[c_{est}\right] = \frac{1}{kn}\left\langle \sigma^2 \mathrm{F}\left(C_{alarm}, \{C_j\}\right)\right\rangle, \qquad (F5)$$

$$\mathrm{F}\left(C_{alarm}, \{C_j\}\right) = 1 + \left(C_{alarm} - \langle C \rangle\right)^2 \frac{\left(C - \langle C \rangle\right)^2}{\left\langle \left(C - \langle C \rangle\right)^2 \right\rangle^2}$$
$$+ 2\left(C_{alarm} - \langle C \rangle\right)\frac{\left(C - \langle C \rangle\right)}{\left\langle \left(C - \langle C \rangle\right)^2 \right\rangle}$$

[AppF13]. As the right-hand side of equation (F5) will be approximated by replacing σ_j by s_j, the distribution of the resulting expression is important. Similar to the Satterthwaite approximation, the following is approximated as chi-square distributed:

$$v_{eff} \frac{\left\langle s^2 \mathrm{F}\left(C_{alarm}, \{C_j\}\right)\right\rangle}{\left\langle \sigma^2 \mathrm{F}\left(C_{alarm}, \{C_j\}\right)\right\rangle} \approx \chi^2_{v_{eff}} \qquad (F6)$$

[AppF14], where the effective number of degrees of freedom v_{eff} is determined so that the variances of both sides of equation (F6) agree:

$$v_{eff} = kv_R \frac{\left\langle \sigma^2 \mathrm{F}\left(C_{alarm}, \{C_j\}\right)\right\rangle^2}{\left\langle \sigma^4 \left[\mathrm{F}\left(C_{alarm}, \{C_j\}\right)\right]^2 \right\rangle} \qquad (F7)$$

Components for Evaluation of Direct Reading Monitors

[AppF15], where v_R is the number of degrees of freedom in each replicate:

$$v_R = n - 1.$$

As in the Satterthwaite approximation, the right-hand side of equation (F7) is approximated by replacing σ_j by s_j.

A noncentral t-variable can now be introduced:

$$t \equiv \frac{\left(c_{est} - c\left[C_{alarm}\right]\right) \big/ \text{var}\left[c_{est}\right]^{1/2} + \lambda}{\sqrt{\chi^2 / v_{eff}}} \qquad (F8)$$

[AppF16]. The noncentrality parameter λ is chosen to approximate a confidence limit on the right-hand side of equation (F4):

$$\lambda = -\sqrt{kn} u_{\alpha_-} \frac{\sigma\left[C_{alarm}\right]}{\left\langle \sigma^2 \, F\left(C_{alarm}, \{C_j\}\right)\right\rangle^{1/2}} \qquad (F9)$$

[AppF17], where the numerator is approximated, using the first regression results, by

$$\sigma\left[C_{alarm}\right] \approx \sqrt{\hat{\sigma}_0^2 + \hat{\sigma}_1^2 C_{alarm}^2},$$

and the denominator, by replacing σ_j by s_j.

If $t_{\gamma, v_{eff}, \lambda}$ is the γ-quantile value

$$c_{alarm} > c_{est}\left[C_{alarm}\right] - \frac{t_{\gamma, v_{eff}, \lambda}}{\sqrt{kn}} \left\langle s^2 \, F\left(C_{alarm}, \{C_j\}\right)\right\rangle^{1/2}$$

[AppF18] at probability $= \gamma$ (e.g., 95%), then, at confidence level γ, the single-sided confidence limit $c_{alarm, \gamma}$ on c_{alarm} is

$$c_{alarm, \gamma} = c_{est}\left[C_{alarm}\right] \\ - \frac{t_{\gamma, v_{eff}, \lambda}}{\sqrt{kn}} \left\langle s^2 \, F\left(C_{alarm}, \{C_j\}\right)\right\rangle^{1/2} \qquad (F10)$$

[AppF19]. To the extent that the approximations used here are accurate, with probability γ, no more than 5% of the population of measurements c_{est} taken at C_{alarm} will be less than $c_{alarm, \gamma}$. The practical implication of this is that the alarm should be turned on at $c_{alarm, \gamma}$. (See simulation results in Section 5 of this appendix.)

As was discussed in Appendix A, the monitor unit's determinations may be corrected for estimated bias by dividing $c_{est}\left[C_{alarm}\right]$ by $(1 + \hat{B})$. This correction requires division of the square root expression in equation (F10) by $(1 + \hat{B})$ and inclusion of an addend for the variance of $(1 + \hat{B})$ inside the square root in equation (F10). This addend depends on the how the estimator of $(1 + B)$ is obtained, as described in Appendix A. If bias is treated as known, then $(1 + B)$ is to be used in the above corrections, and there is no additional addend.

4. Controlling False Positives (σ and c Unknown)

Suppose C_0 is a safe concentration (e.g., 0), where the alarm should stay *off* at probability $(1 - \alpha_+)$. Similar approximations as above result in a confidence limit $c_{+, \gamma}$, below which the alarm is set to *off*. The quantity $c_{+, \gamma}$ equals $c_{alarm, \gamma}$ with C_{alarm} replaced by C_0 and u_{α_-} by $u_{1 - \alpha_+}$. In addition, since an upper confidence limit $c_{+, \gamma}$ is required, $t_{\gamma, v_{eff}, \lambda}$ must be replaced by $t_{1 - \gamma, v_{eff}, \lambda}$ in (F10).

The variability of the system must be sufficiently under control that the false positive limit $c_{+, \gamma}$ is less than the false negative limit $c_{alarm, \gamma}$. For illustration, Figure F–1 indicates mean values of the two limits for a particular variability: $k = 4$, $n = 10$, σ_0 ranges between 0% and 20% of $\langle C \rangle$, and σ_1 is fixed at 10% of $\langle C \rangle$. As is evident, σ_0 must be less than about 20% of $\langle C \rangle$ to attain at least the desired error rates (5%) at confidence level $\gamma = 95\%$ in the evaluation. Here, C_{alarm} is assumed to equal $\langle C \rangle$. The data are assumed to have been corrected for bias.

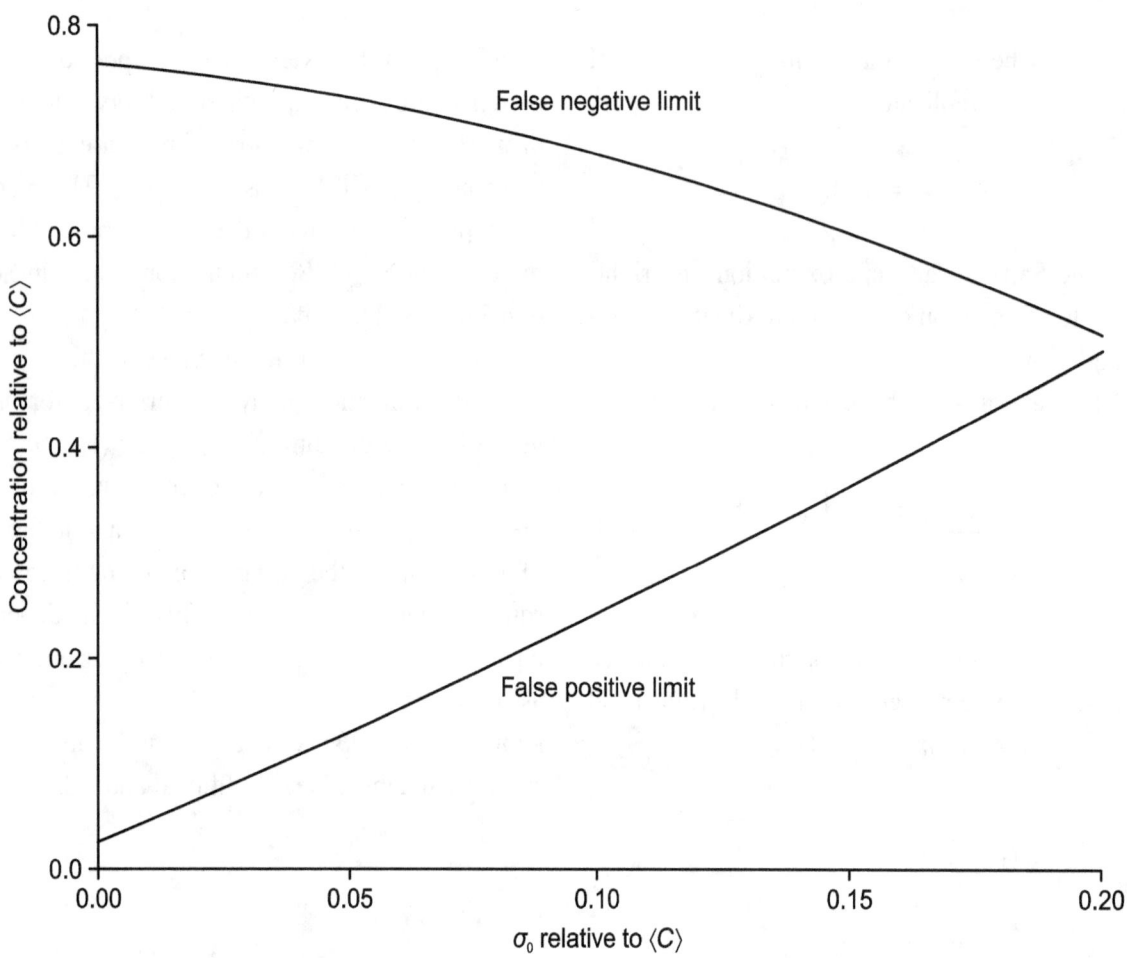

Figure F–1. Mean values for false positive and false negative limits. See text for conditions.

5. Verification

The above algorithms were tested with 5000-point simulations using the computer program presented in Appendix G. The target evaluation confidence γ_{target} = 95%. For each evaluation, k = 4 and n = 10. In the simulations for the following situations there was no correction for bias. The program in Appendix G does not include a bias correction.

False negative tests:

$$C_{alarm} = \langle C \rangle$$
$$\sigma_0 = 20\% \langle C \rangle \text{ and } \sigma_1 = 10\%$$
$$\delta_0 = 10\% \langle C \rangle \text{ and } \delta_1 = 10\%$$
Result: $1 - \gamma = 4.0\%$

$$C_{alarm} = 0.5 \langle C \rangle$$
$$\sigma_0 = 20\% \langle C \rangle \text{ and } \sigma_1 = 10\%$$

$$\delta_0 = 10\% \langle C \rangle \text{ and } \delta_1 = 10\%$$
Result: $1 - \gamma = 6\%$

False positive test at $C_0 = 0$:

$$\sigma_0 = 20\% \langle C \rangle \text{ and } \sigma_1 = 10\%$$
$$\delta_0 = 10\% \langle C \rangle \text{ and } \delta_1 = 10\%$$
Result: $1 - \gamma = 8\%$

This last result is somewhat anticonservative.

5.1. Influence Parameters

The above formalism can be expanded to account for influence parameters [ISO 1993], i.e., variations induced by variation of an environmental parameter (e.g., temperature or humidity) where the instrument output's dependence on the parameter

Components for Evaluation of Direct Reading Monitors

has been measured, and the expected variation in the parameter is specified. Rather than equation (F1), let the instrument response be

$$\hat{c} = c + \hat{\varepsilon} + \hat{\varepsilon}_{ext}$$

[AppF20], where $\hat{\varepsilon}_{ext}$ is a normally distributed external random variable with variance estimate s_{ext}^2.

A conservative confidence limit on c_{alarm} can be obtained as follows. First, note that c_{alarm} is given now by

$$c_{alarm} = c[C_{alarm}] + u_{\alpha_-}\sqrt{\sigma[C_{alarm}]^2 + \sigma_{ext}^2}$$

[AppF21]. A random variable distributed as noncentral t is set up as in equation (F8), except that the denominator is broadened from the expressions in equation (F6) by pooling s_{ext}^2 into the numerator of equation (F6) and σ_{ext}^2 into the denominator and by computing a value for v_{eff} from Satterthwaite's approximation using knowledge of the variability in s_{ext}^2. Then, note that the resulting t is more variable than a quantity t' defined by replacing $\text{var}[c_{est}]^{1/2}$ in

equation (F8) by (F5) with σ_{ext}^2 pooled in. By defining the noncentrality parameter γ, expanding the denominator of equation (F9) as above, and replacing $\sigma[C_{alarm}]$ in the numerator by

$$\sqrt{\sigma[C_{alarm}]^2 + \sigma_{ext}^2},$$

equation (F10) becomes

$$c_{alarm,\gamma} > c_{est}[C_{alarm}]$$
$$+ \frac{t_{\gamma,v_{eff},\lambda}}{\sqrt{kn}} \left\langle s_{ext}^2 + s^2\, F\left(C_{alarm},\{C_j\}\right) \right\rangle^{1/2}$$

[AppF22] as a confidence limit, which is conservative due to the increased variability mentioned above. This expression has not been verified via simulation.

6. References

ISO [1993]. Guide to the expression of uncertainty in measurement. Geneva, Switzerland: International Organization for Standardization.

Appendix G. R Program for Implementing Appendix F— Lower Confidence Limit For Negatives, Controlling False Negatives

The program below is written in the R programming language [R Project 2011]. It can be used either to give estimates for sample data or to simulate one sample.

References

R Project [2011]. The R project for statistical computing [http://www.r-project.org/]. Date accessed: November 2011.

```
# R program for alarm system evaluation with lower confidence limit for
#   negatives, controlling false negatives. It can be used either to give
#   estimates for sample data or to simulate one sample.
Calarm <- 1.0
C0 <- 0.0
# C1 is the vector of concentrations.
C1 <- c(0.0, 0.6666, 1.3333, 2)
Cmean <- mean(C1)
# k is the number of levels.
k <- length(C1)
# n is the number of replications in each level.
n <- 10
nuR <- n - 1
# For the simulations sig0 and sig1, specify the intercept and slope relating
#   variance to concentration-squared.
sig0 <- 0.2
sig1 <- 0.1
# For the simulations del0 and del1, specify the intercept and slope relating
#   measured concentration to true concentration.
del0 <- 0.1
del1 <- 1.1
# Generate a vector of 1s equal to number of replicates at each concentration.
one10 <- matrix(1, n, 1)
C <- C1
Chat <- matrix(0, 4, 3)
C1.mat <- matrix(0, 4, 2)
Chat[, 1] <- C1
C1.mat[, 1] <- C1
# Generate random sample.
# If result is wanted for a data set, set nn <- 1.
nn <- 10000
Cdat.mat <- matrix(0, nn, 4)
Vdat.mat <- matrix(0, nn, 4)
calmlim <- matrix(0, nn, 2)
for(j in (1:nn)) {
  for (i in (1:4)) {
    n1 <- rnorm(n, mean=0, sd=1) * sig0
    n2 <- rnorm(n, mean=0, sd=1) * sig1
    Chat[i, 2] <- mean(del0 * one10 + del1 * C1.mat[i, 1] * one10 +
                  (n1^1 + C1.mat[i, 1]^1 * n2^1) * one10)
    Chat[i, 3] <- var(del0 * one10 + del1 * C1.mat[i, 1] * one10 +
                  (n1^1 + C1.mat[i, 1]^1 * n2^1) * one10)
```

```
}
# Example data means and variances get stored in columns 2 and 3,
#   respectively, of Chat matrix.
# Chat[, 2] <- t(c(0.0244, 0.936, 1.536, 2.367))
# Chat[, 3] <- t(c(0.0386, 0.01899, 0.10576, 0.059))
Cdat.mat[j, ] <- t(Chat[, 2])
Vdat.mat[j, ] <- t(Chat[, 3])
Cal_Cbar <- Calarm - Cmean
var_cbar <- (Chat[, 1])
k <- length(Chat[, 2])
mult <- Cal_Cbar / (var(var_cbar) * (k - 1) / k)
one <- matrix(c(rep(1, k)), k, 1)
dif <- Chat[, 1] - Cmean * one
num <- Chat[, 3] * (one + mult * dif) * (one + mult * dif)
num2 <- (mean(num))^2
den <- mean(Chat[, 3] * Chat[, 3] * (one + mult * dif) *
            (one + mult * dif) * (one + mult * dif) * (one + mult * dif))
# Result below corresponds to equation (F7).
nu_eff <- k * (n - 1) * num2 / den
C2 <- Chat[, 1] * Chat[, 1]
# Steps below estimate intercept and slope for variances.
lm.var <- lm(formula=Chat[, 3] ~ C2)
coef1 <- coef(lm.var)
sig_alarm <- coef1[1] + coef1[2] * Calarm^2
den2 <- (mean(num))^0.5
# Step below corresponds to equation (F9).
lam <- -(k * n)^0.5 * qnorm(0.05) * sig_alarm^0.5 / den2
# Step below estimates intercept and slope for concentrations.
lm.conc <- lm(formula=Chat[, 2] ~ Chat[, 1])
coef2 <- coef(lm.conc)
conc_alarm <- coef2[1] + coef2[2] * Calarm
# Step below corresponds to equation (F10), the lower confidence limit for the
#   concentration at which the instrument should alarm.
c_alarm_gam <- conc_alarm - qt(0.95, nu_eff, lam) / (k * n)^0.5 * den2
calmlim[j, 1] <- j
calmlim[j, 2] <- c_alarm_gam
}
summary(Cdat.mat)
summary(Vdat.mat)
alamtr <- del0 + del1 * Calarm + qnorm(0.05) * (sig0^2 + sig1^2 * Calarm^2)^0.5
onelong <- matrix(1, length(calmlim[, 2]), 1)
id <- ifelse(calmlim[, 2] - alamtr * onelong>0, 1, 0)
# For simulations, the summary below gives the fraction of simulations for which
#   the confidence limit for the alarm value is greater than the alarm.
summary(id)
# The mean below is average value of confidence limit (for simulations) or the
#   actual value for example data.
mean(calmlim[, 2])
```

Appendix H. LaTeX Translations of Selected Mathematical Formulas

Mathematical formulas in this document were prepared using MathType [Design Science 2012] version 6.7a. In Appendices A–F, the more complicated formulas are followed by bracketed reference numbers, e.g., [AppA01]. Table H–1 gives, for each referenced formula, the LaTeX [LaTeX Project 2012] translation using the MathType "LaTeX 2.09 and later" cut and copy preference.

References

Design Science [2012]. MathType 6.7 [http://www.dessci.com/en/products/mathtype/]. Date accessed: February 2012.

LaTeX Project [2012]. LaTeX – A document preparation system [http://www.latex-project.org]. Date accessed: March 2012.

Table H–1. LaTeX translations of selected mathematical formulas

Reference Number	LaTeX Translation
AppA01	`\[\Phi \left( {\frac{{B + A}}{{{S_{{\rm{rT}}}}}}} \right) - \Phi \left( {\frac{{B - A}}{{{S_{{\rm{rT}}}}}}} \right) = 0.95\]`
AppA02	`\[A\left( {B,{S_{{\rm{rT}}}}} \right) = \left\{ \begin{array}{l} 1.96 \times \sqrt {{B^2} + S_{{\rm{rT}}}^2} & {\rm{if }}\left| B \right| < \frac{{{S_{{\rm{rT}}}}}}{{1.645}}{\rm{,}}\\ \left| B \right| + 1.645 \times {S_{{\rm{rT}}}} & {\rm{otherwise}} \end{array} \right.\]`
AppA03	`\[\begin{array}{l} \hat A = A\left({\hat B,{{\hat S}_{{\rm{rT}}}}} \right)\\ = \left\{ \begin{array}{l} 1.96 \times \sqrt {{{\hat B}^2} + \hat S_{{\rm{rT}}}^2} & {\rm{if }} \left| {\hat B} \right| < \frac{{{{\hat S}_{{\rm{rT}}}}}}{{1.645}}{\rm{,}}\\ \left| {\hat B} \right| + 1.645 \times {{\hat S}_{{\rm{rT}}}} & {\rm{otherwise}} \end{array} \right. \end{array}\]`
AppA04	`\[{\hat A_p} = \left\{ \begin{array}{l} 1.96 \times \lambda \times \sqrt {{{\hat B}^2} + \hat S_{{\rm{rT}}}^2} & {\rm{if }}\left| {\hat B} \right| < \frac{{{{\hat S}_{{\rm{rT}}}}}}{{1.645}}{\rm{,}}\\ \left| {\hat B} \right| + 1.645\left(\tau \right){{\hat S}_{{\rm{rT}}}} & {\rm{otherwise}} \end{array} \right.\]`

(Continued)

Reference Number	LaTeX Translation

AppA05	```\[\begin{array}{l}```
	```\lambda = \sqrt {{M \mathord{\left/```
	```{\vphantom {M {\chi _{1 - p,M}^2}}} \right.```
	```\kern-\nulldelimiterspace} {\chi _{1 - p,M}^2}}} ,\\```
	```\tau = {{{t_{p,M}}\left(\Delta \right)} \mathord{\left/```
	```{\vphantom {{{t_{p,M}}\left(\Delta \right)} \Delta }} \right.```
	```\kern-\nulldelimiterspace} \Delta },\\```
	```\Delta = 1.645 \times \sqrt N```
	```\end{array}\]```

AppA06	```\[\begin{array}{l}```
	```\lambda = \sqrt {{\nu \mathord{\left/```
	```{\vphantom {\nu {\chi _{1 - p,\nu }^2}}} \right.```
	```\kern-\nulldelimiterspace} {\chi _{1 - p,\nu }^2}}} {\rm{,}}\\```
	```\nu = {{{{{\left[{{{\hat B}^2} + \hat S_{{\rm{rT}}}^2} \right]}}^2}}```
	```\mathord{\left/```
	```{\vphantom {{{{{\left[{{{\hat B}^2} + \hat S_{{\rm{rT}}}^2}```
	```\right]}}^2}} {\left\{ {\left[{\left({{2 \mathord{\left/```
	```{\vphantom {2 N}} \right.```
	```\kern-\nulldelimiterspace} N}} \right){{\hat B}^2} + {{\hat S_{{```
	```\rm{rT}}}}^2} \mathord{\left/```
	```{\vphantom {{\hat S_{{\rm{rT}}}^2} M}} \right.```
	```\kern-\nulldelimiterspace} M}} \right]\hat S_{{\rm{rT}}}^2 \right\}}}}```
	```\right.```
	```\kern-\nulldelimiterspace} {\left\{ {\left[{\left({{2 \mathord{\left/```
	```{\vphantom {2 N}} \right.```
	```\kern-\nulldelimiterspace} N}} \right){{\hat B}^2} + {{\hat S_{{```
	```\rm{rT}}}}^2} \mathord{\left/```
	```{\vphantom {{\hat S_{{\rm{rT}}}^2} M}} \right.```
	```\kern-\nulldelimiterspace} M}} \right]\hat S_{{\rm{rT}}}^2 \right\}}}}```
	```\end{array}\]```

AppA07	```\[\begin{array}{l}```
	```\hat A = 1.96{u_{\rm{c}}}\\```
	```= 1.96\sqrt {{{{{{\left({1 + {1 \mathord{\left/```
	```{\vphantom {1 s}} \right.```
	```\kern-\nulldelimiterspace} s}} \right)\hat S_{{\rm{rT}}}^2} \mathord```
	```{\left/```
	```{\vphantom {{\left({1 + {1 \mathord{\left/```
	```{\vphantom {1 s}} \right.```
	```\kern-\nulldelimiterspace} s}} \right)\hat S_{{\rm{rT}}}^2} {\left```
	```({1 + \hat B} \right)}}}} \right.```
	```\kern-\nulldelimiterspace} {\left({1 + \hat B} \right)}}}^2} +```
	```\hat S_{{{\rm{r}}}{{\rm{T}}}_{{\rm{Ref,Avg}}}}^2} \\```
	```= 1.96\sqrt {\left({1 + {1 \mathord{\left/```
	```{\vphantom {1 s}} \right.```
	```\kern-\nulldelimiterspace} s}} \right)\hat S_{\rm{r}}^2 + \hat S_{{```
	```\rm{r}}{{\rm{T}}_{{\rm{Ref,Avg}}}}}^2}```
	```\end{array}\]```

(Continued)

Table H–1 (Continued). LaTeX translations of selected mathematical formulas

Reference Number	LaTeX Translation
AppA08	`\[{A_p} = \hat A\sqrt {\frac{{\frac{\nu }{{\chi _{1 - p,\nu }^2}}}}{{\left( {1 + \frac{2}{{{s^2}}} + \frac{{2\hat S_{{\rm{rT,Ref,Avg}}}^4}}{{S_{\rm{r}}^4}} + \frac{5}{s}\frac{{\hat S_{{\rm{rT,Ref,Avg}}}^2}}{{S_{\rm{r}}^2}}} \right)}}} \]`

AppA09	`\[\nu = \frac{{u_{\rm{c}}^4}}{{\left\{ \begin{array}{l}`
	`\hat S_{{\rm{rT,Ref,Avg}}}^4 + \frac{2}{s}\hat S_{{\rm{rT,Ref,Avg}}}^2`
	`\hat S_{\rm{r}}^2\\`
	` + \left[ {\frac{1}{{s - 1}}{{\left( {1 + \frac{1}{s}} \right)}^2} + `
	`\frac{1}{{{s^2}}}} \right]\hat S_{\rm{r}}^4`
	`\end{array} \right\}}}\]`

AppA10	`\[\begin{array}{l}`
	`\hat A = 1.96{u_{\rm{c}}}\\`
	` = 1.96\sqrt {\frac{1}{3}\Delta _{\max }^2 + \left( {1 + \frac{1}{s}} \right)\hat S_{\rm{r}}^2}`
	`\end{array}\]`

AppA11	`\[{A_p} = \hat A\sqrt {\frac{{\frac{\nu }{{\chi _{1 - p,\nu }^2}}}}{{\left( {1 + \frac{2}{{{s^2}}} + \frac{{\frac{4}{{45}}\Delta _{\max }^4}}{{S_{\rm{r}}^4}} + \frac{5}{s}\frac{{\frac{1}{3}\Delta _{\max }^2}}{{S_{\rm{r}}^2}}} \right)}}} \]`

AppA12	`\[\nu = \frac{{u_{\rm{c}}^4}}{{\left\{ \begin{array}{l}`
	`\frac{2}{{45}}\Delta _{\max }^4 + \frac{2}{{3s}}\Delta _{\max }^2\hat S_{\rm{r}}^2\\`
	` + \left[ {\frac{1}{{s - 1}}{{\left( {1 + \frac{1}{s}} \right)}^2} + \frac{1}{{{s^2}}}} \right]\hat S_{\rm{r}}^4`
	`\end{array} \right\}}}\]`

AppB01	`\[\begin{array}{l}`
	`{{\mathop{\rm SS}\nolimits} _w}\left( y \right) = \sum\limits_{i = 1}^k {\sum\limits_{j = 1}^n {{{\left( {{y_{ij}} - {{\bar y}_{i \cdot }}} \right)}^2}} ,} \\`
	`{{\bar y}_{i \cdot }} = \frac{1}{n}\sum\limits_{j = 1}^n {{y_{ij}}} {\rm{, }}{y_{ij}} = \frac{{{x_{ij}}}}{{{C_{{\rm{T}}i}}}} - 1`
	`\end{array}\]`

AppB02	`\[{{\mathop{\rm SS}\nolimits} _b}\left( y \right) = \sum\limits_{i = 1}^k {n{{\left( {{{{\bar y}_{i \cdot }} - {{\bar y}_{ \cdot \cdot }}} \right)}^2}} {\rm{, }}{\bar y_{ \cdot \cdot }} = \frac{1}{k}\sum\limits_{i = 1}^k {{{\bar y}_{i \cdot }}} \]`
AppB03	`\[F = \frac{{{{{{{\mathop{\rm SS}\nolimits} }_b}\left(y \right)} \mathord{\left/ {\vphantom {{{{{\mathop{\rm SS}\nolimits} }_b}\left(y \right)} {\left({k - 1} \right)}}} \right. \kern-\nulldelimiterspace} {\left({k - 1} \right)}}}}}{{{{{{{\mathop{\rm SS}\nolimits} }_w}\left(y \right)} \mathord{\left/ {\vphantom {{{{{\mathop{\rm SS}\nolimits} }_w}\left(y \right)} {\left({k\left({n - 1} \right)} \right)}}} \right. \kern-\nulldelimiterspace} {\left({k\left({n - 1} \right)} \right)}}}}}\]`

(Continued)

Table H–1 (Continued). AMS-LaTeX translations of selected mathematical formulas

Reference Number	LaTeX Translation

AppB04	`\[F' = \frac{{{{{{{\mathop{\rm SS}\nolimits} }_b}\left( {y'} \right)} \mathord{\left/`
	 `{\vphantom {{{{{{\mathop{\rm SS}\nolimits} }_b}\left( {y'} \right)} {\left( {k - 1} \right)}}}} \right.`
	 `\kern-\nulldelimiterspace} {\left( {k - 1} \right)}}}}}{{{{{{{{\mathop`
	 `{\rm SS}\nolimits} }_w}\left( {y'} \right)} \mathord{\left/`
	 `{\vphantom {{{{{\mathop{\rm SS}\nolimits} }_w}\left( {y'} \right)}`
	 `{\left( {k\left( {n - 1} \right)} \right)}}}} \right.`
	 `\kern-\nulldelimiterspace} {\left( {k\left( {n - 1} \right)} \right)}}}`
	 `+ \left( {{n \mathord{\left/`
	 `{\vphantom {n m}} \right.`
	 `\kern-\nulldelimiterspace} m}} \right)\hat \sigma _{\rm{R}}^2}}\]`

AppB05	`\[\begin{array}{l}`
	 `{{\hat C}_{{\rm{T}}i}} = \frac{1}{m}\sum\limits_{j = 1}^m {{c_{ij}}}`
	 `{\rm{, and also }}\\`
	 `\hat \sigma _{\rm{R}}^2 = \frac{1}{{k\left( {m - 1} \right)}}\sum`
	 `\limits_{i = 1}^k {\sum\limits_{j = 1}^m {{{{{\left( {{c_{ij}} -`
	 `{{\hat C}_{{\rm{T}}i}}} \right)}^2}} \mathord{\left/`
	 `{\vphantom {{{{\left( {{c_{ij}} - {{\hat C}_{{\rm{T}}i}}} \right)}^2}}`
	 `{\hat C_{{\rm{T}}i}^2}}} \right.`
	 `\kern-\nulldelimiterspace} {\hat C_{{\rm{T}}i}^2}}} }`
	 `\end{array}\]`

AppB06	`\[\hat \mu  = \frac{{\sum\limits_{i = 1}^k {\left( {{z_{i1}`
	 `{\rm{(test)}}}} + m{{\bar z}_{i \cdot {\rm{(ind)}}}}} \right)} }}`
	 `{{\left( {1 + m} \right)k}}{\rm{, }}{\hat \mu _{{{\rm{test}}}} = \frac`
	 `{{\sum\limits_{i = 1}^k {{z_{i1{\rm{(test)}}}}} }}{k}\]`

AppB07	`\[{\hat \mu _{{{\rm{ind}}}}} = \frac{{\sum\limits_{i = 1}^k {m{{\bar z}_{i`
	 `\cdot {\rm{(ind)}}}}} }}{{mk}}{\rm{, }}{\hat \mu _i} = \frac{{{z_{i1`
	 `{\rm{(test)}}}} + m{{\bar z}_{i \cdot {\rm{(ind)}}}}}}{{1 + m}}\]`

AppB08	`\[{\hat \alpha _{{{\rm{test}}}}} = {\hat \mu _{{{\rm{test}}}}} - \hat \mu`
	 `{\rm{, }}{\hat \alpha _{{{\rm{ind}}}}} = {\hat \mu _{{{\rm{ind}}}}} - \hat`
	 `\mu {\rm{, }}{\hat \beta _i} = {\hat \mu _i} - \hat \mu \]`

AppB09	`\[{\hat \sigma _{{{\rm{bias}}}}} = \sqrt {\frac{{\left( \begin{array}{l}`
	 `\sum\limits_{i = 1}^k {z_{i1{\rm{(test)}}}^2}  + \sum\limits_{i = 1}^k`
	 `{m\bar z_{i \cdot {\rm{(ind)}}}^2} \\`
	 ` - \left[ \begin{array}{l}`
	 `k\left( {1 + m} \right){{\hat \mu }^2} + k\left( {\hat \alpha _{{`
	 `\rm{test}}}^2 + m\hat \alpha _{{{\rm{ind}}}^2} \right)\\`
	 ` + \sum\limits_{i = 1}^k {\left( {1 + m} \right)\hat \beta _i^2}`
	 `\end{array} \right]`
	 `\end{array} \right)}}{{{{\left( {k - 1} \right)}}}} \]`

AppB10	`\[{\hat \sigma _{{{\rm{res}}}}} = \sqrt {\frac{{\sum\limits_{i = 1}^k {{{`
	 `\left[ {{z_{ij{\rm{(ind)}}}}} - {{\bar z}_{i \cdot {\rm{(ind)}}}}}`
	 `\right]}^2}} }}{{k\left( {m - 1} \right)}}} \]`

(Continued)

Reference Number	LaTeX Translation
AppB11	`\[\hat \sigma _{{\rm{test}}}^2 = \frac{{\left( {k - 1} \right)\hat \sigma _{{\rm{bias}}}^2 + k\left( {m - 1} \right)\hat \sigma _{{ \rm{res}}}^2}}{{km - 1}}\]`
AppB12	`\[{\bar x_i} = \sum\limits_{j = 1}^n {{{{x_{ij}}}} \mathord{\left/ {\vphantom {{{x_{ij}}} n}} \right. \kern-\nulldelimiterspace} n}} \]`
AppB13	`\[H = \frac{{k\left({n - 1} \right)\ln \left({\hat \sigma _{\rm{e}}^2} \right) - \sum\limits_{i = 1}^k {\left({n - 1} \right)\ln \left({\hat \sigma _i^2} \right)} }}{{1 + {{\left({k + 1} \right)} \mathord{\left/ {\vphantom {{\left({k + 1} \right)} {\left[{3k\left({n - 1} \right)} \right]}}} \right. \kern-\nulldelimiterspace} {\left[{3k\left({n - 1} \right)} \right]}}}}}\]`
AppB14	`\[\hat \sigma _{\rm{e}}^{\rm{2}} = \frac{1}{k}\sum\limits_{i = 1}^k {\hat \sigma _i^{\rm{2}}} \]`
AppB15	`\[\hat \sigma _i^{\rm{2}} = \frac{1}{{n - 1}}\sum\limits_{j = 1}^n {{{{\left( {{y_{ij}} - {{\bar y}_i}} \right)}^2}} \]`
AppB16	`\[H = \frac{{N'\ln \left( {\hat \sigma _{\rm{e}}^2} \right) - \sum\limits_{i = 1}^k {\left( {{n_i} - 1} \right)\ln \left( {\hat \sigma _i^2} \right)} }}{{1 + \frac{1}{{3\left( {k - 1} \right)}}\left[ {\left( {\sum\limits_{i = 1}^k {\frac{1}{{{n_i} - 1}}} } \right) - \frac{1}{{N'}}} \right]}}\]`
AppB17	`\[\hat \sigma _{\rm{e}}^2 = \sum\limits_{i = 1}^k {\left( {{n_i} - 1} \right)\frac{{\hat \sigma _i^2}}{{N'}}} \]`
AppB18	`\[N' = \sum\limits_{i = 1}^k {\left( {{n_i} - 1} \right)} \]`
AppB19	`\[{\hat S_{{\rm{rT}}}} = \sqrt {\sum\limits_{i = 1}^k {\frac{{\left( {{n_i} - 1} \right) \times {{\left( {{{\hat S}_{{\rm{rT}}i}}} \right)}^2}}}{{\sum\limits_{l = 1}^k {\left( {{n_l} - 1} \right)} }}} } \]`
AppB20	`\[{\hat S_{\rm{r}}} = \sqrt {\sum\limits_{i = 1}^k {\frac{{\left( {{n_i} - 1} \right) \times {{\left( {{{\hat S}_{{\rm{r}}i}}} \right)}^2}}}{{\sum\limits_{l = 1}^k {\left( {{n_l} - 1} \right)} }}} } \]`

(Continued)

Table H–1 (Continued). AMS-LaTeX translations of selected mathematical formulas

Reference Number	LaTeX Translation
AppB21	`\[{\nu _{{\rm{test,alt}}}} = \frac{{{{\left\{ {\left[{{{(m + 1)} \mathord{\left/ {\vphantom {{(m + 1)} m}} \right. \kern-\nulldelimiterspace} m}} \right]\hat \sigma _{{\rm{bias}}}}^2 - {{\hat \sigma _{{\rm{res}}}}^2} \mathord{\left/ {\vphantom {{\hat \sigma _{{\rm{res}}}}^2 m}} \right. \kern-\nulldelimiterspace} m}} \right\}}}^2}}}{{\frac{{{{\left\{ {\left[{{{\left({m + 1} \right)} \mathord{\left/ {\vphantom {{\left({m + 1} \right)} m}} \right. \kern-\nulldelimiterspace} m}} \right]\hat \sigma _{{\rm{bias}}}}^2} \right\}}}^2}}}{{\left({k - 1} \right)}} + \frac{{{{\left({{{\hat \sigma _{{\rm{res}}}}^2} \mathord{\left/ {\vphantom {{\hat \sigma _{{\rm{res}}}}^2 m}} \right. \kern-\nulldelimiterspace} m}} \right)}^2}}}{{k\left({m - 1} \right)}}}}\]`
AppB22	`\[\begin{array}{l} {\mathop{\rm MSE}\nolimits} \left( x \right) = \frac{1}{{d\left( {n - 1} \right)}}\sum\limits_{i = 1}^d {\sum\limits_{j = 1}^n {{{\left( {{x_{ij}} - {{\bar x}_{i \cdot }}} \right)}^2}} } ,\\ {{\bar x}_{i \cdot }} = \frac{1}{n}\sum\limits_{j = 1}^n {{x_{ij}}} \end{array}\]`
AppB23	`\[\begin{array}{l} {\mathop{\rm MSB}\nolimits} \left( x \right) = \frac{1}{{d - 1}}\sum\limits_{i = 1}^d {n{{\left( {{{\bar x}_{i \cdot }} - {{\bar x}_{ \cdot \cdot }}} \right)}^2}} ,\\ {{\bar x}_{ \cdot \cdot }} = \frac{1}{d}\sum\limits_{i = 1}^d {{{\bar x}_{i \cdot }}} \end{array}\]`
AppB24	`\[\begin{array}{l} {{\hat S}_{{\rm{rTW}}}} = \sqrt {\hat S_{{\rm{rTE}}}^2 + \hat S_{{\rm{rTD}}}^2} \\ = \sqrt {\left( {1 - \frac{1}{n}} \right) \times {\mathop{\rm MSE}\nolimits} \left( x \right) + \frac{1}{n} \times {\mathop{\rm MSB}\nolimits} \left( x \right)} \end{array}\]`
AppB25	`\[{\nu _1} = \frac{{{{{\left[{\left({1 - \frac{1}{n}} \right) \times {\mathop{\rm MSE}\nolimits} \left(x \right) + \frac{1}{n} \times {\mathop{\rm MSB}\nolimits} \left(x \right)} \right]}^2}}}{{\frac{{{{\left({1 - \frac{1}{n}} \right) \times {{\left[{{\mathop{\rm MSE}\nolimits} \left(x \right)} \right]}^2}}}}{{dn}} + \frac{{{{{\left[{{\mathop{\rm MSB}\nolimits} \left(x \right)} \right]}^2}}}}{{\left({d - 1} \right){n^2}}}}}}\]`

(Continued)

Reference Number	LaTeX Translation
AppB26	`\[{{\nu _{\rm{D}}} = \frac{{\hat S_{{\rm{rD}}}}^4}}{{\frac{{{{\left[{\left({{1 \mathord{\left/ {\vphantom {1 n}} \right. \kern-\nulldelimiterspace} n}} \right){\mathop{\rm MSE}\nolimits} \left(z \right)} \right]}^2}}}{{d\left({n - 1} \right)}} + \frac{{{{\left[{\left({{1 \mathord{\left/ {\vphantom {1 n}} \right. \kern-\nulldelimiterspace} n}} \right){\mathop{\rm MSB}\nolimits} \left(z \right)} \right]}^2}}}{{d - 1}}}}}\]`
AppB27	`\[\frac{{{{{\left( {\hat S_{{\rm{rD}}}}^2 + \hat \sigma _{{\rm{test}}}}^2} \right)}^2}}}{{\frac{{\hat S_{{\rm{rD}}}}^4}}{{{{\nu _{\rm{D}}}}}} + \frac{{\hat \sigma _{{\rm{test}}}}^4}}{{km - 1}}}}}\]`
AppB28	`\[\frac{{{{{\left( {\hat S_{{\rm{rD}}}}^2 + \hat \sigma _{{\rm{test,alt}}}}^2} \right)}^2}}}{{\frac{{\hat S_{{\rm{rD}}}}^4}}{{{\nu _{\rm{D}}}}}} + \frac{{\hat \sigma _{{\rm{test,alt}}}}^4}}{{{\nu _{{\rm{test,alt}}}}}}}}}}\]`
AppC01	`\[\begin{array}{c} {{\mathop{\rm MS}\nolimits} _{\rm{U}}}\left(y \right) = \frac{{kd\sum\limits_{i = 1}^u {{{\left({{{\bar y}_{i \cdot \cdot }} - {{\bar y}_{\cdot \cdot \cdot }}} \right)}^2}} }}{{u - 1}},\\ {{\bar y}_{i \cdot \cdot }} = \sum\limits_{j,l} {{{{y_{ijl}}} \mathord{\left/ {\vphantom {{{{y_{ijl}}} {\left({kd} \right)}}} \right. \kern-\nulldelimiterspace} {\left({kd} \right)}}} ,\\ {{\bar y}_{ \cdot \cdot \cdot }} = \sum\limits_{i,j,l} {{{{y_{ijl}}} \mathord{\left/ {\vphantom {{{{y_{ijl}}} {\left({kdu} \right)}}} \right. \kern-\nulldelimiterspace} {\left({kdu} \right)}}} \end{array}\]`
AppC02	`\[\begin{array}{c} {{\mathop{\rm MS}\nolimits} _{\rm{D}}}\left(y \right) = \frac{{ku\sum\limits_{j = 1}^d {{{\left({{{\bar y}_{ \cdot j \cdot }} - {{\bar y}_{ \cdot \cdot \cdot }}} \right)}^2}} }}{{d - 1}},\\ {{\bar y}_{ \cdot j \cdot }} = \sum\limits_{i,l} {{{{y_{il}}} \mathord{\left/ {\vphantom {{{{y_{il}}} {\left({ku} \right)}}} \right. \kern-\nulldelimiterspace} {\left({ku} \right)}}} \end{array}\]`

(Continued)

Reference Number	LaTeX Translation

AppC03	`\[\begin{array}{c}`
	`{{\mathop{\rm MS}\nolimits} _{{\rm{UD}}}}}\left( y \right) = \frac{{k`
	`\sum\limits_{i,j} {{{\left( {{{\bar y}_{ij \cdot }} - {{\bar y}_{i`
	`\cdot \cdot }} - {{\bar y}_{ \cdot j \cdot }} + {{\bar y}_{ \cdot`
	`\cdot \cdot }}} \right)}^2}} }}{{\left( {u - 1} \right)\left( {d - 1}`
	`\right)}},\\`
	`{{\bar y}_{ij \cdot }} = \sum\limits_{l = 1}^k {{{{y_{ijl}}} \mathord{`
	`\left/`
	`{\vphantom {{{y_{ijl}}} k}} \right.`
	`\kern-\nulldelimiterspace} k}}`
	`\end{array}\]`

AppC04	`\[\begin{array}{c}`
	`{{\mathop{\rm MS}\nolimits} _{{\rm{tD}}}}}\left( y \right) = \frac{{`
	`\left[ \begin{array}{l}`
	`u\sum\limits_{j,l} {{{\left( {{{\bar y}_{ \cdot jl}} - {{\bar y}_{ \cdot`
	`\cdot \cdot }}} \right)}^2}} \\`
	` - \left( {d - 1} \right){{\mathop{\rm MS}\nolimits} _{\rm{D}}}}\left( y`
	`\right)`
	`\end{array} \right]}}{{d\left( {k - 1} \right)}},\\`
	`{{\bar y}_{ \cdot jl}} = \sum\limits_{i = 1}^u {{{{y_{ijl}}} \mathord{`
	`\left/`
	`{\vphantom {{{y_{ijl}}} u}} \right.`
	`\kern-\nulldelimiterspace} u}}`
	`\end{array}\]`

AppC05	`\[{{\mathop{\rm MS}\nolimits} _{{\rm{res}}}}}\left( y \right) = \frac{{`
	`\left[ \begin{array}{l}`
	`\sum\limits_{i,j,l} {\bar y_{ijl}^2} - \left( {kdu} \right)\bar y_{`
	`\cdot \cdot \cdot }^2\\`
	` & - \left( {u - 1} \right){{\mathop{\rm MS}\nolimits} _{\rm{U}}}}\left`
	`( y \right)\\`
	` & - \left( {d - 1} \right){{\mathop{\rm MS}\nolimits} _{\rm{D}}}}\left`
	`( y \right)\\`
	` & - \left( {u - 1} \right)\left( {d - 1} \right){{\mathop{\rm MS}`
	`\nolimits} _{{\rm{UD}}}}}\left( y \right)\\`
	` & - d\left( {k - 1} \right){{\mathop{\rm MS}\nolimits} _{{\rm{tD}}}}}`
	`\left( y \right)`
	`\end{array} \right]}}{{d\left( {u - 1} \right)\left( {k - 1} \right)}}\]`

AppC06	`\[\sum\limits_{i,l} { \equiv \sum\limits_{i = 1}^u {\sum\limits_{l =` `1}^k {} } } \]`
AppC07	`\[\hat B = \exp \left( {{{\bar z}_{ \cdot \cdot \cdot {\rm{(test)}}}}}` ` - {{\bar z}_{ \cdot \cdot \cdot ({\rm{ind}})}}}}} \right)\]`
AppC08	`\[{{\hat S_{{\rm{rTB}}}}} = \sqrt {\frac{{{{\mathop{\rm MSB}\nolimits}` `\left( y \right) - {\mathop{\rm MSE}\nolimits} \left( y \right)}}{k}} \]`

(Continued)

Reference Number	LaTeX Translation
AppC09	`\[{\bar y_{i \cdot }} = \frac{1}{k}\sum\limits_{l = 1}^k {{y_{il}}} {` `\rm{, }}{\bar y_{ \cdot l}} = \frac{1}{u}\sum\limits_{i = 1}^u {{y_` `{il}}} {\rm{, }}{\bar y_{ \cdot  \cdot }} = \frac{1}{u}\sum\limits_{i =` `1}^u {{{\bar y}_{i \cdot }}} \]`
AppC10	`\[\begin{array}{c}` `{\mathop{\rm MSE}\nolimits} \left( y \right) = {1 \mathord{\left/` `{\vphantom {1 {\left[ {\left( {k - 1} \right)\left( {u - 1} \right)}` `\right]}}}} \right.` `\kern-\nulldelimiterspace} {\left[ {\left( {k - 1} \right)\left( {u -` `1} \right)} \right]}}\\` `\times \sum\limits_{i = 1}^u {\sum\limits_{l = 1}^k {{{\left( {{y_{il}}` `- {{\bar y}_{i \cdot }} - {{\bar y}_{ \cdot l}} + {{\bar y}_{ \cdot` `\cdot }}} \right)}^2}} }` `\end{array}\]`
AppC11	`\[{\mathop{\rm MSB}\nolimits} \left( y \right) = \frac{1}{{u - 1}}\sum` `\limits_{i = 1}^u {k{{\left( {{{\bar y}_{i \cdot }} - {{\bar y}_{ \cdot` `\cdot }}} \right)}^2}} \]`
AppC12	`\[{\nu _2} = \frac{{{{\left\{ {{{\left[ {{\mathop{\rm MSB}\nolimits}` `\left( y \right) - {\mathop{\rm MSE}\nolimits} \left( y \right)}` `\right]} \mathord{\left/` `{\vphantom {{\left[ {{\mathop{\rm MSB}\nolimits} \left( y \right) - {` `\mathop{\rm MSE}\nolimits} \left( y \right)} \right]} k}} \right.` `\kern-\nulldelimiterspace} k}} \right\}}^2}}}{{\frac{{{{\left[ {\left` `( {{1 \mathord{\left/` `{\vphantom {1 k}} \right.` `\kern-\nulldelimiterspace} k}} \right)}{\mathop{\rm MSB}\nolimits}` `\left( y \right)} \right]}}^2}}}{{u - 1}} + \frac{{{{\left[ {\left( {{1` `\mathord{\left/` `{\vphantom {1 k}} \right.` `\kern-\nulldelimiterspace} k}} \right)}{\mathop{\rm MSE}\nolimits}` `\left( y \right)} \right]}}^2}}}{{\left( {u - 1} \right)\left( {k - 1}` `\right)}}}}\]`
AppC13	`\[{\hat S_{{\rm{rT}}}} = \sqrt {\hat S_{{\rm{rTW}}}^2 + \hat S_{{` `\rm{rTB}}}^2} \]`
AppC14	`\[\begin{array}{c}` `\hat S_{{\rm{rT}}}^2 = \frac{{k - 1}}{k}{{\mathop{\rm MS}\nolimits} _{{` `\rm{res}}}}\left( y \right) + \frac{{d - 1}}{{dk}}{{\mathop{\rm MS}` `\nolimits} _{{\rm{UD}}}}\left( y \right)\\` `+ \frac{1}{{dk}}{{\mathop{\rm MS}\nolimits} _{\rm{U}}}\left( y \right)` `\end{array}\]`

(Continued)

Reference Number	LaTeX Translation

AppC15	`\[\frac{{\hat S_{{\rm{rT}}}}^4}}{{\left( \begin{array}{l}`
	 `\frac{{{{\left[ {\left( {1 - {1 \mathord{\left/`
	 `{\vphantom {1 k}} \right.`
	 `\kern-\nulldelimiterspace} k}} \right)}{{\mathop{\rm MS}\nolimits} }_`
	 `{{\rm{res}}}}}\left( y \right)} \right]}^2}}}{{d\left( {k - 1} \right)`
	 `\left( {u - 1} \right)}}\\`
	 `& + \frac{{{{\left[ {\left( {{1 \mathord{\left/`
	 `{\vphantom {1 k}} \right.`
	 `\kern-\nulldelimiterspace} k} - {1 \mathord{\left/`
	 `{\vphantom {1 {\left( {dk} \right)}}} \right.`
	 `\kern-\nulldelimiterspace} {\left( {dk} \right)}}} \right)}{{{\mathop{`
	 `\rm MS}\nolimits} }_{{\rm{UD}}}}}\left( y \right)} \right]}^2}}}{{\left`
	 `( {d - 1} \right)\left( {u - 1} \right)}}\\`
	 `& + \frac{{{{\left[ {\left( {{1 \mathord{\left/`
	 `{\vphantom {1 {\left( {dk} \right)}}} \right.`
	 `\kern-\nulldelimiterspace} {\left( {dk} \right)}}} \right)}{{{\mathop{`
	 `\rm MS}\nolimits} }_{\rm{U}}}}\left( y \right)} \right]}^2}}}{{\left( {u`
	 `- 1} \right)}}`
	 `\end{array} \right)}}\]`

AppD01	`\[\left( {\frac{y}{{1 + ku}}{\rm{, }}\frac{y}{{1 - ku}}} \right)\]`
AppD02	`\[k = {t_{(1 + p)/2,\nu }}\]`
AppD03	`\[k = {t_{(1 + p)/2,\nu }}\]`

AppD04	`\[\begin{array}{l}`
	 `A = 1.960 \times {S_{{\rm{rT}}}}\\`
	 `\doteq 1.960 \times \sqrt {{\nu \mathord{\left/`
	 `{\vphantom {\nu  {\chi _{0.05,\nu }^2}}} \right.`
	 `\kern-\nulldelimiterspace} {\chi _{0.05,\nu }^2}}}  \times {{\hat S}_`
	 `{{\rm{rT}}}}`
	 `\end{array}\]`

AppD05	`$\chi _{0.05,\nu }^2$`

AppD06	`\[\begin{array}{l}`
	 `k = 1.960 \times \sqrt {{{15} \mathord{\left/`
	 `{\vphantom {{15} {\chi _{0.05,15}^2}}} \right.`
	 `\kern-\nulldelimiterspace} {\chi _{0.05,15}^2}}} \\`
	 `= 2.8`
	 `\end{array}\]`

AppF01	`\[\hat c = c + \hat \varepsilon \]`
AppF02	`\[c = {\delta _0} + \left( {1 + {\delta _1}} \right)C\]`
AppF03	`\[{\sigma ^2} = \sigma _0^2 + \sigma _1^2{C^2}\]`

(Continued)

Reference Number	LaTeX Translation
AppF04	`\[\frac{1}{{\sqrt {2\pi \sigma } }}\int\limits_{ - \infty }^{{c_{{ \rm{alarm}}}}} {{e^{{{ - {\textstyle{1 \over 2}}{{(\hat c - c)}^2}} \mathord{\left/ {\vphantom {{ - {\textstyle{1 \over 2}}{{(\hat c - c)}^2}} {{\sigma ^2}}}} \right. \kern-\nulldelimiterspace} {{\sigma ^2}}}}}{\mathop{\rm d}\nolimits} \hat c} = {\alpha _ - }\]`
AppF05	`\[{c_{{\rm{alarm}}}} = c\left[ {{C_{{\rm{alarm}}}}} \right] + {u_{{ \alpha _ - }}}\sigma \left[ {{C_{{\rm{alarm}}}}} \right]\]`
AppF06	`\[\hat s_j^2 = \hat \sigma _0^2 + \hat \sigma _1^2C_j^2\]`
AppF07	`\[{\bar c_j} = {\hat \delta _0} + \left( {1 + {{\hat \delta }_1}} \right){C_j}\]`
AppF08	`\[{\mathop{\rm var}} \left[ {{c_{{\rm{est}}}}} \right] = {\mathop{ \rm var}} \left[ {\left\langle {\bar c} \right\rangle  + \left( {{C_{{ \rm{alarm}}}} - \left\langle C \right\rangle } \right){{\hat \delta }_1}} \right]\]`
AppF09	`\[{\mathop{\rm var}} \left[ {\left\langle {\bar c} \right\rangle } \right] = \frac{1}{{kn}}\left\langle {{\sigma ^2}} \right\rangle \]`
AppF10	`\[{\mathop{\rm var}} \left[{\left\langle {{{\hat \delta }_1}} \right \rangle } \right] = \frac{1}{{kn}}\frac{{\left\langle {{\sigma ^2}{{ \left({C - \left\langle C \right\rangle } \right)}^2}} \right\rangle }} {{{{\left\langle {{{\left({C - \left\langle C \right\rangle } \right)}^2}} \right\rangle }^2}}}\]`
AppF11	`\[{\mathop{\rm cov}} \left[{\left\langle {\bar c} \right\rangle ,{{\hat \delta }_1}} \right] = \frac{1}{{kn}}\frac{{\left\langle {{\sigma ^2} \left({C - \left\langle C \right\rangle } \right)} \right\rangle }}{{ \left\langle {{{\left({C - \left\langle C \right\rangle } \right)}^2}} \right\rangle }}\]`
AppF12	`\[\begin{array}{l} {\mathop{\rm var}} \left[{\left\langle {\bar c} \right\rangle + \left ({{C_{{\rm{alarm}}}} - \left\langle C \right\rangle } \right)\left({1 + {{\hat \delta }_1}} \right)} \right] = \\ & {\mathop{\rm var}} \left[{\left\langle {\bar c} \right\rangle } \right] + {\left({{C_{{\rm{alarm}}}} - \left\langle C \right\rangle } \right)^2}{\mathop{\rm var}} \left[{{{\hat \delta }_1}} \right]\\ & & + 2\left({{C_{{\rm{alarm}}}} - \left\langle C \right\rangle } \right){\mathop{\rm cov}} \left({\left\langle {\bar c} \right\rangle ,{{\hat \delta }_1}} \right) \end{array}\]`

(Continued)

Table H–1 (Continued). AMS-LaTeX translations of selected mathematical formulas

Reference Number	LaTeX Translation

AppF13	`\[\begin{array}{l}`
	`{\mathop{\rm var}} \left[ {{c_{{\rm{est}}}}} \right] = \frac{1}{{kn}}`
	`\left\langle {{\sigma ^2}{\mathop{\rm F}\nolimits} \left( {{C_{{`
	`\rm{alarm}}}},\left\{ {{C_j}} \right\}} \right)} \right\rangle ,\\`
	`{\mathop{\rm F}\nolimits} \left( {{C_{{\rm{alarm}}}},\left\{ {{C_j}}`
	`\right\}} \right) = 1 + {\left( {{C_{{\rm{alarm}}}} - \left\langle C`
	`\right\rangle } \right)^2}\frac{{{{\left( {C - \left\langle C \right`
	`\rangle } \right)}^2}}}{{{{\left\langle {{{\left( {C - \left\langle C`
	`\right\rangle } \right)}^2}} \right\rangle }^2}}}\\`
	`& + 2\left( {{C_{{\rm{alarm}}}} - \left\langle C \right\rangle }`
	`\right)\frac{{\left( {C - \left\langle C \right\rangle } \right)}}{{`
	`\left\langle {{{\left( {C - \left\langle C \right\rangle } \right)}^2}}`
	`\right\rangle }}`
	`\end{array}\]`

AppF14	`\[{\nu _{{\rm{eff}}}}\frac{{\left\langle {{s^2}{\mathop{\rm F}\nolimits}`
	`\left( {{C_{{\rm{alarm}}}},\left\{ {{C_j}} \right\}} \right)} \right`
	`\rangle }}{{\left\langle {{\sigma ^2}{\mathop{\rm F}\nolimits} \left`
	`( {{C_{{\rm{alarm}}}},\left\{ {{C_j}} \right\}} \right)} \right\rangle`
	`}} \approx \chi _{{\nu _{{\rm{eff}}}}}^2\]`

AppF15	`\[{\nu _{{\rm{eff}}}} = k{\nu _{\rm{R}}}\frac{{{{\left\langle {{\sigma`
	`^2}{\mathop{\rm F}\nolimits} \left( {{C_{{\rm{alarm}}}},\left\{ {{C_j}}`
	`\right\}} \right)} \right\rangle }^2}}}{{\left\langle {{\sigma ^4}{{`
	`\left[ {{\mathop{\rm F}\nolimits} \left( {{C_{{\rm{alarm}}}},\left`
	`\{ {{C_j}} \right\}} \right)} \right]}^2}} \right\rangle }}\]`

AppF16	`\[t \equiv \frac{{{{\left( {{c_{{\rm{est}}}}} - c\left[ {{C_{{`
	`\rm{alarm}}}}} \right]} \right)} \mathord{\left/`
	`{\vphantom {{\left( {{c_{{\rm{est}}}}} - c\left[ {{C_{{\rm{alarm}}}}}`
	`\right]} \right)} {{\mathop{\rm var}} {{\left[ {{c_{{\rm{est}}}}}}`
	`\right]}^{1/2}} + \lambda }}} \right.`
	`\kern-\nulldelimiterspace} {{\mathop{\rm var}} {{\left[ {{c_{{`
	`\rm{est}}}}} \right]}^{1/2}} + \lambda }}}}{{\sqrt {{{{\chi ^2}}`
	`\mathord{\left/`
	`{\vphantom {{{\chi ^2}} {{\nu _{{\rm{eff}}}}}}}} \right.`
	`\kern-\nulldelimiterspace} {{\nu _{{\rm{eff}}}}}}} }}\]`

AppF17	`\[\lambda = - \sqrt {kn} {u_{\alpha \_}}\frac{{\sigma \left[ {{C_{{`
	`\rm{alarm}}}}} \right]}}{{{{{\left\langle {{\sigma ^2}{\mathop{\rm F}`
	`\nolimits} \left( {{C_{{\rm{alarm}}}},\left\{ {{C_j}} \right\}} \right)}`
	`\right\rangle }^{1/2}}}}\]`

AppF18	`\[{c_{{\rm{alarm}}}} > {c_{{\rm{est}}}}[{C_{{\rm{alarm}}}}] - \frac{{{t_`
	`{\gamma ,{\nu _{{\rm{eff}}}},\lambda }}}}{{\sqrt {kn} }}{\left\langle`
	`{{s^2}{\mathop{\rm F}\nolimits} \left( {{C_{{\rm{alarm}}}},\left`
	`\{ {{C_j}} \right\}} \right)} \right\rangle ^{1/2}}\]`

(Continued)

Reference Number	LaTeX Translation

AppF19	`\[\begin{array}{l}`
	`{c_{{\rm{alarm,}}}\gamma }} = {c_{{\rm{est}}}}\left[ {{C_{{\rm{alarm}}}}} \right]\\`
	`- \frac{{{t_{\gamma ,{\nu _{{\rm{eff}}}}},\lambda }}}}{{\sqrt {kn} }}{`
	`\left\langle {{s^2}{\mathop{\rm F}\nolimits} \left( {{C_{{`
	`\rm{alarm}}}},\left\{ {{C_j}} \right\}} \right)} \right\rangle ^{1/2}}`
	`\end{array}\]`

AppF20	`\[\hat c = c + \hat \varepsilon  + {\hat \varepsilon _{{\rm{ext}}}}\]`

AppF21	`\[{{c_{{\rm{alarm}}}}} = c\left[ {{C_{{\rm{alarm}}}}} \right] + {u_{{`
	`\alpha _ - }}}\sqrt {\sigma {{\left[ {{C_{{\rm{alarm}}}}} \right]}}^2} +`
	`\sigma _{{\rm{ext}}}}^2} \]`

AppF22	`\[\begin{array}{l}`
	`{c_{{\rm{alarm}},\gamma }} > {c_{{\rm{est}}}}}\left[ {{C_{{\rm{alarm}}}}}`
	`\right]\\`
	`+ \frac{{{t_{\gamma ,{\nu _{{\rm{eff}}}}},\lambda }}}}{{\sqrt {kn} }}`
	`{\left\langle {s_{{\rm{ext}}}}^2 + {s^2}{\mathop{\rm F}\nolimits} \left`
	`( {{C_{{\rm{alarm}}}}},\left\{ {{C_j}} \right\}} \right)} \right\rangle`
	`^{1/2}}`
	`\end{array}\]`

Bibliography

Bartley DL [2001]. Definition and assessment of sampling and analytical accuracy. Ann Occup Hyg *45*(5):357–364.

Hald A [1952]. Statistical theory with engineering applications. New York: John Wiley & Sons.

ISO [1993]. Guide to the expression of uncertainty in measurement. Geneva, Switzerland: International Organization for Standardization.

Johnson NL, Kotz S [1970]. Continuous univariate distributions. Vols. 1–2. Boston, MA: Houghton Mifflin Company.

Johnson NL, Leone FC [1964]. Statistics and experimental design. Vol. 1. New York: John Wiley & Sons.

Kenny LC, Lidén G [1993]. The application of performance standards to personal airborne dust samplers. Ann Occup Hyg *33*(3):289–300.

NIOSH [1977]. SCP statistical protocol. By Busch KA. In: Taylor DG, Kupel RE, Bryant JM, eds. Documentation of the NIOSH validation tests. Cincinnati, OH: U.S. Department of Health, Education, and Welfare, Center for Disease Control, National Institute for Occupational Safety and Health, DHEW (NIOSH) Publication No. 77–185, NTIS No. PB-274248, pp. 7–12.

NIOSH [1980]. Development and validation of methods for sampling and analysis of workplace toxic substances. By Gunderson EC, Anderson CC. Cincinnati, OH: U.S. Department of Health and Human Services, Centers for Disease Control, National Institute for Occupational Safety and Health, DHHS (NIOSH) Publication No. 80–133, NTIS No. PB89–182042.

NIOSH [1994]. NIOSH manual of analytical methods (NMAM™). 4th ed. Eller PM, Cassinelli ME, O'Connor PF, Schlecht PC, eds. Cincinnati, OH: U.S. Department of Health and Human Services, Centers for Disease Control and Prevention, National Institute for Occupational Safety and Health, DHHS (NIOSH) Publication No. 94–113; 1st Supplement, Publication No. 96–135; 2nd Supplement, Publication No. 98–119; 3rd Supplement, Publication No. 2003–154. [http://www.cdc.gov/niosh/nmam/]

NIOSH [1995]. Guidelines for air sampling and analytical method development and evaluation. By Kennedy ER, Fischbach TJ, Song R, Eller PM, Shulman SA. Cincinnati, OH: U.S. Department of Health and Human Services, Centers for Disease Control and Prevention, National Institute for Occupational Safety and Health, DHHS (NIOSH) Publication No. 95–117. [http://www.cdc.gov/niosh/docs/95-117/pdfs/95-117.pdf]

R Project [2011]. The R project for statistical computing [http://www.r-project.org/]. Date accessed: November 2011.

Satterthwaite FE [1946]. An approximate distribution of estimates of variance components. Biometrics Bull *2*(6):110–114.

Smith H [1936]. The problem of comparing the results of two experiments with unequal errors. J Counc Sci Ind Res *9*:211–212.

Wald A [1942]. Setting of tolerance limits when the sample is large. Ann Math Stat *13*(4):389–399.

Wald A [1943]. An extension of Wilks' method for setting tolerance limits. Ann Math Stat *14*(1):45–55.

Wald A, Wolfowitz J [1946]. Tolerance limits for a normal distribution. Ann Math Stat *17*(2):208–215.

Wilks SS [1941]. Determination of sample sizes for setting tolerance limits. Ann Math Stat *12*(1):91–96.

Wilks SS [1942]. Statistical prediction with special reference to the problem of tolerance limits. Ann Math Stat *13*(4):400–409.

www.ingramcontent.com/pod-product-compliance
Lightning Source LLC
Chambersburg PA
CBHW081503170526
45166CB00008B/2532